T0265768

SCIENCE IN AN AGE OF UNREASON

Science
in an Age *of*
Unreason

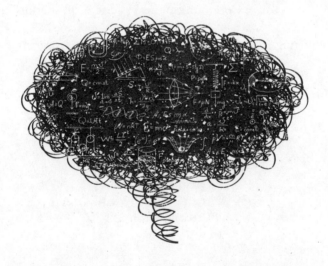

John Staddon

REGNERY GATEWAY
Washington, D.C.

Regnery Gateway™ is a trademark of Salem Communications Holding Corporation
Regnery® is a registered trademark and its colophon is a trademark of Salem Communications Holding Corporation

Cataloging-in-Publication data on file with the Library of Congress

ISBN: 978-1-68451-252-2
eISBN: 978-1-68451-323-9
Library of Congress Control Number: 2021949799

Published in the United States by
Regnery Gateway, an Imprint of
Regnery Publishing
A Division of Salem Media Group
Washington, D.C.
www.RegneryGateway.com

Manufactured in the United States of America

10 9 8 7 6 5 4 3 2 1

Books are available in quantity for promotional or premium use. For information on discounts and terms, please visit our website: www.Regnery.com.

Contents

Preface

S cience is in trouble. Some of the problems are internal—the replication crisis, a surplus of scientists, fissiparous subdivision of journals and professional organizations, and what that means for critical review. Other problems reflect issues in the wider society. Facts can cause people to react emotionally. Sometimes that's appropriate: Finding a gas leak in the basement of your building should cause alarm and make you flee and warn your neighbors. It is an emergency. But what about the following claim (from a letter to a college-alumni magazine): "I don't think racism is totally responsible for the plight of minority victims"?

David Hume, star of the Scottish Enlightenment, made a simple distinction, vital for science, between *facts* and *wishes,* between *is* and *ought*.[1] Science is about facts; discerning wishes, or what *ought* to be done, is a purview of other disciplines. There are facts, or factual claims, in both these examples. In the first case, the facts are indubitable, there is a leak; it is an emergency, action is required. But in the second case, there is no emergency, and the facts, unlike the gas leak, are not self-evident. Reason—science—demands that the claims be verified: Are family arrangements a problem for the success of poor black people? Do these communities have an unproductive attitude toward work and education? Only if the claims are true is some action justified.

Yet the immediate reaction to these comments was not inquiry but condemnation: reflexive cries of "racism."[2] Now society reacts this way to many findings and areas of research, not just race and gender, but also climate change. Topics must not be studied, or only studied with a foregone conclusion in mind. Fact versus passion, but all too often passion wins. This tendency threatens the integrity of science, social science especially. A purpose of this book is to clear up muddle and, perhaps, stem the tide.

Science has other external problems: the job market for scientists, changes in motivation when science shifts from being a vocation to a career, funding patterns and misaligned incentives. These factors have turned many academic departments away from scholarship and toward political bias, if not outright activism. These political factors interact with internal problems to weaken and distort the findings of science. I look at the effects on sociology and history of science.

Many scientific questions give rise to what physicist Alvin Weinberg—in an almost-forgotten article—called *trans-science*,[3] by which he meant questions that are scientific but, for practical or ethical reasons, cannot be answered conclusively by the methods of science. Examples are the small, long-delayed effects of low-concentration pollutants, the causes of climate change, and the role of genes in intelligence. When decisive science is impossible, other factors dominate. Weak science lets slip the dogs of unreason: many social scientists have difficulty separating *facts* from *faith*, reality from the way they would like things to be. Critical research topics have become taboo which, in turn, means that policy makers are making decisions based more on ideologically driven political pressure than scientific fact.

The book is in five parts. Part 1 is mostly philosophy. It deals with science and faith in the context of Darwinian evolution. Many secular humanists and some evolutionary biologists believe that science provides an ethical system—not the long-discredited "social Darwinism," but a mélange of supposedly secular values derived from liberal and progressive writing over the past three centuries. This issue has implications for the way that religion is treated by U.S. law. Chapter 1 lays out the problem. The next two chapters discuss different aspects of this issue. Chapter 4 summarizes what we actually know of Darwinian evolution.

Part 2 discusses contemporary problems with science as a profession. Are there too many scientists? Is the problem lack of jobs or a shortage of solvable problems? How have the incentives for science changed over the decades? Why is scientific publishing still in the nineteenth century? One challenge is the injection of social-justice ideology into government science funding bodies such as the National Science Foundation. This growing trend devalues merit as a criterion for support and diverts resources from science itself.

Part 3 is on a contentious contemporary issue: climate change. There is a consensus that human-produced greenhouse gases are warming the atmosphere to a dangerous degree. Much passion is involved: critics are termed "deniers," and apocalyptic consequences are predicted if massive, society-transforming steps are not taken. President Biden calls climate change an "existential threat."[4] Yet the evidence is largely circumstantial, and the consequences are probably less than catastrophic and possibly even benign. It is another trans-science problem.

Part 4 is on social science, which has fragmented to the point that more than one hundred disciplines and subdisciplines study human social behavior, with different vocabularies and critical standards in an increasingly politicized climate. Most sociological problems are in the trans-science category: real experiments are impossible, and correlations are regularly morphed into causes. The technical issues involved in the most-frequently-used method in the social and bio-medical sciences are discussed in the appendix.

Adding to the intrinsic difficulty of social science, race, particularly, has become a topic where disinterested research on the causes of racial disparities, for example, is almost impossible. "Scientific" conclusions increasingly reflect ideological predispositions, rather than appropriately cautious inferences from necessarily inadequate data. Hence the rise of the influential concept of systemic racism. Systemic racism is unmeasurable, hence ineradicable. Its rise has been accompanied by a stifling of research that might shed real light on racial and gender disparities: the study of individual differences in ability and interest. This suppression bears an uncomfortable resemblance to the tragedy of Soviet Lysenkoism (Chapter 14).

Part 5 is on history of science, which is subject to two ailments. The first is understanding: many historians have little interest in technical matters, and more in social issues and personalities. Yet to understand history of science one must first understand the science. Second is politics: there is a growing tendency to interpret scientific findings in terms not of the science, but of an imagined ideology either of scientists or their audience. This section discusses a range of biases: a gifted scientist with frank political bias who has also written a good history of science; professional historians who seem to know the science but interpret everything in political terms; and two other well-known historians, persuasive writers both, with a superficial understanding of the social science about which they write. The result is not so much history of science as political journalism, or docudrama of the "inspired-by-real-events" type. The book ends with an afterword touching on missed topics.

I'm sure I have come to some wrong conclusions. I hope that others will write to correct them so we can all move a little closer to a very complex truth.

This book is necessarily a personal view. The scope of the problem and my own interests and abilities limit what could be covered. Nevertheless, I hope it presents a reasonable sample of the current problems of science, a noble activity whose integrity is essential to the survival of our civilization.

PART 1

Evolution

CHAPTER 1

Has Secular Humanism Made Science a Religion?

What Is Religion?

I s secular humanism a religion? In 1995 the U.S. Court of Appeals
for the Ninth Circuit examined the issue and concluded, rightly,
that science, in the form of the theory of evolution, is *not* a religion.[1] On the other hand, in 2006, the BBC aired an excellent program called *The Trouble with Atheism* which argued that atheists are
religious and made the point via a series of interviews with prominent atheists who claimed their beliefs were "proved" by science.[2] The
presenter, feisty journalist Rod Liddle, concluded that Darwinism is
in effect a religion.

Liddle may be correct as far as atheist scientists are concerned:
many of them do indeed speak with religious fervor about their
beliefs.[3] But he is wrong about science itself. As eighteenth century
philosopher David Hume showed many years ago, science consists
of facts, but facts alone do not motivate. Without *motive*, a fact points
to no action. Liddle was right, however, in this: both religion and

secular humanism *do* provide motives, explicit in one case, but covert in the other.

The Elements of Religion

All religions have three elements, although the relative emphasis differs from one religion to another.[4]

The first is a belief in invisible or hidden beings, worlds, and processes—like God, angels, heaven, miracles, reincarnation, and the soul. All are unseen and unseeable, except by mystics under special and generally unrepeatable conditions. All are unverifiable by the methods of science. Hence, from a scientific point of view, these features of religion are neither true nor false, but simply unprovable. They have no implications for action. Since we don't as yet have laws restricting thought, these beliefs should have no bearing on legal matters.

The second element is claims about the real world. Every religion, especially in its primordial version, makes claims that are essentially scientific—assertions of fact that are potentially verifiable. These claims are of two kinds. The first we might call timeless: for example, claims about physical nature—the geocentric solar system, the Hindu turtle that supports the world, properties of foods, the doctrine of literal transubstantiation. The second are claims about history: Noah's flood, the age of the earth, the resurrection—all "myths of origin." Some of these claims are unverifiable; as for the rest, there is a consensus that science usually wins—in law and elsewhere. In any case, few of these claims have any bearing on action. Like the first category, they are ideas and not commands.

The third property of a religion is its rules for action, its morality. All religions have a code, a set of moral and behavioral prescriptions, matters of belief—usually, but not necessarily, said to flow from

God—that provide guides to action in a wide range of situations: the Ten Commandments, the principles of Sharia, the Five Precepts of Buddhism and Jainism, et cetera.

Neo-Christian

Secular humanism denies the supernatural and defers matters of fact to science. But it is as rich in moral rules, in dogma, as any religion.[5]

Its rules come neither from God nor from reason, but from texts like John Stuart Mill's *On Liberty*, and from the works of philosophers Baruch Spinoza, Jean-Jacques Rousseau, Peter Singer, Dan Dennett, and John Rawls, psychologists such as B. F. Skinner and Sigmund Freud, and public intellectuals like Sam Harris, Richard Dawkins, and the late Christopher Hitchens. Where *these* folk got their not-always-consistent morals is a matter of dispute, but Judeo-Christian teaching is not uninvolved: consider "the last shall be first and the first last" (Matthew 20:16) and those favored "marginalized groups," for example. (Jesus's next line, "For many are called but few are chosen," which seems to recognize the inevitability of inequality, has been dropped, apparently.) In terms of *moral rules*, secular humanism is indistinguishable from a religion.

It has escaped the kind of attacks directed at Christianity and other up-front religions for two reasons: its name states that it is not religious, and its principles cannot be tracked down to a canonical text. They exist but are not formally defined by any "holy book." But it is only the morality of a religion, not its supernatural or historical beliefs, that has any implications for action, for politics and law. Secular humanism makes moral claims as strong as any other faith. It is therefore as much *religion* as any other. But because it is not seen as religious, the beliefs of secular humanists increasingly influence U.S. law.

The covert nature of these principles is a disadvantage in some ways, but a great advantage in the political/legal context. Because secular-humanist morals cannot be easily identified, they cannot be easily attacked. A secular judgeship candidate can claim to be unbiased, not because she has no religious principles, but because her principles are not obvious. Yet belief in the innocence of abortion or the value of homosexuality, the "normality" of the LGBTQ+ community, or the essential sameness of men and women, may be no less passionate, no less based on faith—no less unprovable—than the opposite beliefs of many frankly religious people.

Secular Morals: Three Examples

Paradoxically, as the marriage rate declined and the rate of cohabitation increased, the legalization of *same-sex* marriage became a hot topic. It was once a minority position among American citizens and their elected representatives,[6] but dwindling opposition led to swift legalization of gay marriage in 2015.[7]

This *bouleversement* changed the meaning of the word *marriage* and introduced unnecessary uncertainty into both social and sexual intercourse.[8] Why did this happen, given the declining importance of marriage itself, the availability of civil-partnership contracts, and the historical opposition of all major religions? We cannot be certain, but two things seem to be important. The first is a secular-humanist commandment as powerful as any of the familiar ten: *the primacy of personal passions*, loosely justified by John Stuart Mill's *harm* principle—you can do whatever you want so long as you don't harm others.[9] The second, alluded to earlier, is a mutation of Christian morals: the "last-shall-be-first" principle. The last-shall-be-first principle demands that *all inequality be rectified*. The different status of same- and different-sex liaisons, and the social awkwardness of the new meaning of the word *marriage*, is dwarfed by these principles, to which secular elites now seem committed.

Secular humanists also have blasphemy rules. Dressing in blackface as a teenager or saying the N-word,[10] even in an educational context, can lead to severe retribution.[11] The speaker's intention is essentially irrelevant, a violation of the *mens rea* principle in law. Virginia Governor Ralph Northam was urged to resign over a decades-old blackface incident. But Connecticut Senator Richard Blumenthal survived what many would consider a more serious sin: exaggerating his military experience.[12] Young Northam committed a single instance of racist blasphemy, while Blumenthal persisted in a lie.

The blasphemy of the N-word is related to the hegemony of individualism through the definition of *harm*. Harm is defined by the "victim," not by the intent of the speaker. Hence "microaggressions," often innocent comments and questions,[13] are minor blasphemies because they upset a hearer in a "marginalized group." The "microaggression" charge also allows the marginalized "victim" to silence the speaker, in accordance with the last-shall-be-first principle.

A final example is the forty-foot-tall Bladensburg (Maryland) cross,[14] erected in 1925 with private money but on public land to commemorate soldiers who died in World War I. Fred Edwords, a former official of the American Humanist Association, was one of the plaintiffs who sought to get the cross declared illegal. "This cross sends a message of Christian favoritism and exclusion of all others," says Mr. Edwords[15]—not that anyone else is excluded from erecting their own monument. Evidently toleration is not one of the secular humanist commandments, but Christianity as anathema is. It seems to be the faith of a competitor that Fred objects to.

Religiously affiliated candidates for public office are often quizzed about their religious beliefs.[16] This is both unfair and largely irrelevant. Whether a candidate believes in transubstantiation or the divinity of Allah has no bearing at all on how he or she will judge the rights of litigants. Beliefs about religious stories and transcendental matters do not guide action.

What matters is the person's moral beliefs—whatever their source—and the person's willingness to disregard them if they conflict with the constitution. Secular candidates have just as many "unprovable beliefs" as religious candidates. The only difference is that secular morality is not written down in a single identifiable source. It is not easily accessible.

Candidates, both religious and nonreligious, should all be subject to the same range of questions—questions not about their religion but about what might be called their "action rules." What should be prohibited? What should be encouraged? In short, what are their "goods" and "bads" and how would they act if their beliefs are in conflict with settled law?

The point is to understand the moral beliefs of the candidate and how he or she is prepared to reconcile them with the law, not his or her adherence to a recognized faith. As it is, many passionate, "religious" beliefs of secular candidates go undetected and unquestioned. Thus, they become law by stealth.[17]

And yet, the central issue remains undebated. Can we deduce morality from science? Secular humanists, by insisting that they are not adherents to a religion, claim the mantle of science to justify their moral beliefs. But the separation between science and faith is long-settled philosophy. Today, given the dominance of covert morality masquerading as science, we need reminding that science is a map not a destination. It tells us what is, not what ought to be.

CHAPTER 2

Science and Faith: Can Morality Be Deduced from the Facts of Science?

"Reason is, and ought only to be the slave of
the passions . . ."

—*David Hume*

S cience occasionally evokes an almost religious allegiance. As
we saw in the previous chapter, a few scientists believe that
science can provide us with rules for living, with a morality.
Here is another example from the wonderful ant biologist and chron-
icler of sociobiology, the late E. O. Wilson:

> If the empiricist world view is correct, *ought* is just short-
> hand for one kind of factual statement, a word that denotes
> what society first chose (or was coerced) to do, and then
> codified.[1]

In a 2009 interview, Wilson added:

> One by one, the great questions of philosophy, including
> "Who are we?" and "Where did we come from?" are being
> answered to different degrees of solidity. So gradually, sci-
> ence is simply taking over the big questions created by

philosophy. Philosophy consists largely of the history of failed models of the brain.[2]

Morality, indeed, everything worth believing, *can* be deduced from science, according to Ed Wilson. Yet this is a claim that flatly contradicts a compelling conclusion of Enlightenment philosophy.

Scientific Imperialism

Wilson is not alone; his confidence in the omnipotence of science, his belief in *scientific imperialism*,[3] is shared by vocal members of the (now not-so-new) New Atheists.[4] Richard Dawkins, a splendid science writer who has nevertheless become a convert to his own nonreligion, says that belief in anything that cannot be scientifically proved, that is, *faith*, "is one of the world's great evils, comparable to the smallpox virus but harder to eradicate."[5] But the New Atheists are not themselves lacking in faith. Indeed, they have a zealous and deeply held faith of their own.

Dawkins deems faith "evil precisely because it requires no justification and brooks no argument."[6] But everyone believes some things they cannot prove scientifically. Many of these things Dawkins would surely allow as irrefutably *good*—the Golden Rule, the virtues of generosity, kindness, compassion, and so on. For many people, these virtues require no spiritual sanction; they are just *right*. Yet they cannot be scientifically proved without circularity. You either believe them or you don't. Dawkins surely does not object to unquestioning belief—*faith*—in those moral precepts.

Atheists have moral beliefs that (it is claimed) are not religiously inspired. It is beliefs that have a religious basis—belief in a God, in the specifics of religious stories, and in moral injunctions derived from scripture—that irk many atheists. They tend to object especially when religious prescriptions and proscriptions violate the (often-unstated) beliefs of twenty-first century *bien-pensants*, such as the special rights of some "marginalized" groups, identity of the sexes, pansexual freedom, the innocence of abortion, the evil of corporal punishment, and other implications of the "last-shall-be-first" commandment.

Feelings

When brave enough to expose their moral beliefs to the light of day, secular humanists often ground their ethical precepts in human feeling. In his book *The Moral Landscape*, New Atheist Sam Harris confronts the issue of science and morality and concludes that science can indeed determine moral values.[7] He solves the ethical problem by arguing that "Values are a certain kind of fact,"[8] while also holding that "questions about values—about meaning, morality, and life's larger purpose—are really questions about the well-being of conscious creatures."[9] He also points to felt experience as a signal of value:

> There's no notion, no version of human morality and human values that I've ever come across that is not at some point reducible to a concern about conscious experience and its possible changes.[10]

Harris's claim is indeed a scientific hypothesis. It could be tested by surveys. I don't think it is true—there are surely some aspects of morality that make no references to an individual's "conscious experience." But even if it were true, it is a *fallacy to assume that if we know the process by which a moral belief arises, we also know whether to accept it or not. A process is just a fact.* In other words, even if everyone agrees to define things that contribute to "human flourishing" as "good," we are not forced by reason alone to agree either on what we mean by "flourishing" or even on this definition of "good." In science, most true claims achieve consensus but, obviously, the converse is false: true facts usually achieve a consensus, but a consensus may not be true.

"Concern about conscious experience"—human feelings—is integral to Sam Harris's scientific take on human morality. But feelings are not a reliable guide to truth, moral or otherwise, if only because many scientific, value-free, statements nevertheless elicit strong emotional reactions. For example, in the *Origin of Species*—which is subtitled *The Preservation of Favoured Races in the Struggle for Life*—Charles Darwin describes many examples of competition: "the more vigorous . . . gradually kill the less vigorous," et cetera.[11] One critic of Darwin has reacted emotionally to the word *struggle* and the implication that some individuals and races survive at the expense of others. The idea that some animal, plant, and human varieties—races—are "superior," in the sense that they will prevail in the "struggle," makes Darwin "obviously racist" in the eyes of this author.[12] Indeed, any reference to human individual differences, especially in relation to "race," will elicit passionate feelings in many readers, no matter how "scientific" the context or disinterested the account. So while facts are neutral, the human reaction to them very often is not. People find it very difficult indeed to separate the factual from the emotional.

This is why philosopher David Hume separated "ought," the dictates of morality, from "is," the facts of science. Reason is value-neutral, Hume argued:

It is not contrary to reason to prefer the destruction of the whole world to the scratching of my finger. It is not contrary to reason for me to chuse my total ruin, to prevent the least uneasiness of an Indian or person wholly unknown to me.[13]

Reason doesn't tell you what to do, it tells you how to do it; it is a map, not a destination, as I said earlier. It is the link between passion (will, motivation) and action: "Reason is, and ought only to be the slave of the passions, and can never pretend to any other office than to serve and obey them."[14] Without passion, the facts established by reason are impotent. The findings of science are neither moral nor immoral, according to Hume. Hume's distinction between "is" and "ought" is not a distinction between doing science and doing religion. It is a distinction between *existing* and *acting*.

"Science-Based" Ethics: Human Flourishing

The crux of Harris's argument, and the basis for much of secular-humanist morality, is the argument that moral values are implied by the pursuit of science. As Harris puts it:

> The very idea of "objective" knowledge (i.e., knowledge acquired through honest observation and reasoning) has values built into it, as every effort we make to discuss facts depends upon principles that we must first value (e.g., logical consistency, reliance on evidence, parsimony, etc.).[15]

That the *pursuit* of science involves values, is of course, correct. Scientists have aims, to discover something or solve a problem, and a love of the activity itself. Darwin loved to collect beetles; Claude Bernard certainly enjoyed the results of vivisection, if not the process.

In short, as Hume argued, any *action* other than a reflex requires some kind of motivation, some kind of *value*. Hence, the fact that *doing* science requires scientists to believe in "logical consistency, reliance on evidence, parsimony, etc." (as Harris says), not to mention honesty and curiosity, does not invalidate Hume. Neither does the fact that pursuing science requires faith in a fixed, hence discoverable, nature.[16] The stability of natural law is not self-evident, like a syllogism or simple arithmetic. In order to seek, a scientist must believe (or at least "act as if he believes") there is something to be found.

Yes, doing science requires values; but the facts thus obtained are not themselves values. The facts that men are on average taller than women, or that tests show that African Americans have lower average IQ than white Americans, are equally value neutral. But human nature being what it is, the second fact is likely to elicit much stronger emotions than the first, even though both are just facts. Neither one impels us to action, unless we feel, as a value, that some race or gender differences are a bad thing.

So where do values come from? Do they come from anywhere? Philosopher Daniel Dennett, who has written at length on the subject, concludes in a slightly exasperated tone:

> If "ought" cannot be derived from "is," just what *can* "ought" be derived from? . . . ethics must be *somehow* based on an appreciation of human nature—on a sense of what a human being is or might be, and on what a human being might want to have or want to be. If *that* is naturalism, then naturalism is no fallacy.[17]

If we are asking a scientific question, then of course Dennett is right. But, again, knowing the source of an ethical belief doesn't mean we should act on it or consider it an imperative. Morality may derive from our nature, but not all instincts are good.

CHAPTER 3

Science and Faith:
Darwin to the Rescue?

S cience cannot provide us with an ethics, but some naturalistic suggestions are more plausible than others. Perhaps the most convincing source of human morality is evolution and natural selection. E. O. Wilson and radical behaviorist B. F. Skinner have both suggested that evolutionary epistemology[1] in some form allows *is* to be transformed into *ought*. Others are more cautious, simply affirming that the emotions that underlie morality derive from our evolution. There is, nevertheless, a rough consensus among atheists that what we think good (or bad) is a product of our evolutionary history. There is no unanimity on which evolutionary impulses are *really* good and bad.[2] (And not all "natural" human emotions are good.) There isn't even unanimity on how evolution itself, Darwin's variation and selection, actually works.

The Naturalistic Fallacy

Despite these caveats, there is an influential school that thinks science can define morality. For example, in his provocatively titled

1971 bestseller *Beyond Freedom and Dignity*, radical behaviorist B. F. Skinner (1904–90) said:

> Questions of this sort . . . are said . . . to involve "value judgments"—to raise questions . . . not about what man *can* do but about what he *ought* to do. It is usually implied that the answers are out of the reach of science. . . . It would be a mistake for the behavioral scientist to agree.[3]

The hypothesis that what *ought* to be (in the moral sense) can be inferred from what *is* was termed the *naturalistic fallacy* by English philosopher G. E. Moore (1873–1958).[4] Obviously, Skinner did not believe it to be fallacy. E. O. Wilson's point is more subtle. He believed that if we understand enough about human evolution that understanding will somehow guide us to a true, universal ethics. "The empiricist argument, then, is that by exploring the biological roots of moral behavior, and explaining their material origins and biases, we should be able to fashion a wiser and more enduring ethical consensus than has gone before."[5]

But Hume's point remains. Again there is a confusion between process and outcome: understanding the historical process that led to a belief can justify a scientific claim, but not a moral one.

Behaviorist Skinner defines "the good" in two ways. One is merely descriptive: "good" and "bad" are just whatever society "reinforces"— rewards or punishes. Not much universal there. His other definition echoes Wilson's: "The ultimate sources [of values] are to be found in the evolution of the species and the evolution of the culture."[6] Perhaps "survival" is a value everyone can agree on. The problem is deciding just what will promote survival and what will endanger it. If "survival" is to be our guide, we must be able to predict, at least in broad outline, the course of biological and cultural evolution.

The assumption that evolutionary history is predictable is closely related to the doctrine of *historicism*, espoused most famously by Karl Marx. It was convincingly criticized by Karl Popper, who wrote: "Marx may be excused for holding the mistaken belief that there is a 'natural law of historical development'; for some of the best scientists of his time . . . believed in the possibility of discovering a *law of evolution*. But there can be no empirical 'law of evolution.'"[7]

There are also practical difficulties. First, looking to "survival" for answers to ethical questions will often point to conclusions that conflict with values that are now deeply held. Are we to abandon them? Second, there are very many cultural and genetic "fitness" questions that simply cannot be decided at all: one problem with "survival" as a value is that in many cases it provides little or no practical guidance.

The Survival *Criterion: What Should We Believe?*

A few examples should suffice to show that deciding on evolutionary "good" and "bad" is as difficult as predicting stock movements a hundred years in the future. For example, alcohol is a poison. Hence, cultures that use alcohol must be less "fit" (in the Darwinian sense) than cultures that do not. But are they? There might be hidden benefits to one or the other that we cannot now foresee. The Puritan consensus was that alcohol was an unmitigated evil. The social benefits associated with moderate drinking were assumed to be outweighed by its bad effects. Yet drinking wine and beer is common to the majority of cultures, and now it turns out that there might even be health benefits to moderate drinking, so the evolutionary balance sheet on alcohol is not yet closed.

Here is another example: alcohol might be controversial, but smoking is certainly bad—isn't it? This is not so clear either. Some

smokers (but by no means all[8]) die from lung cancer and emphysema,
usually in unpleasant ways, which is unquestionably bad for "human
flourishing" as well as individual survival. However, smoking-induced
illnesses generally do not kill until their victims reach their fifties and
sixties, after their productive life is almost over and before they
become a burden to their children and to society. It is an evolutionary truism that life history is determined by adaptive considerations,[9]
and a short but productive life is often "fitter," in a natural-selection
sense, than a longer and less productive one.

Perhaps a society that encourages smoking—which yields a
generally short but productive life—will be more successful in the
long run than one that discourages smoking and has to put up with
a lot of unproductive, old people? The much-touted "Greatest Generation" were mostly smokers, after all.[10] Should we perhaps encourage smoking? There are some data to support the idea. Several
studieshave shown that the lifetime health-care costs for smokers
are lower than for nonsmokers (public-health rhetoric to the contrary) because they tend to die more quickly and collect less pen-
sion money than nonsmokers.[11] Whether or not reduced financial
cost corresponds to evolutionary advantage is of course not known,
but an inverse relation between cost and "fitness" is perhaps more
likely than not.

Argument from evolutionary survival very quickly comes up
against many traditional beliefs. Even obvious virtues like safety, care
for the elderly, and the emancipation of women, not to mention tolerance for anti-progenitive sexual abnormalities, might be questioned by a thoroughgoing evolutionary ethicist. Is it really adaptive
to outfit three-year-olds on tricycles with crash helmets so they grow
up timid and unadventurous, or to fit our cars with air bags and seat
belts so that the reckless and inept are protected from the consequences of their actions? And does it make evolutionary sense to

encourage the brightest young women to delay, and thus limit, child-birth so they can spend the prime of their reproductive lives as engineers and investment bankers rather than mothers? Lee Kuan Yew, president of Singapore, thought a few years ago that it did not. He was pilloried as a eugenicist for providing maternal incentives to well-educated women.[12] But surely a conscientious evolutionary ethicist should applaud him.

The problem of what *really* conduces to "fitness"—of a culture or a race—has become especially acute with advances in medicine. Should parents be allowed to control the sex and other character-istics of their children? Should human cloning be permitted? (It has already happened with eight or nine other mammals.) What extraordinary measures are justified to keep a sick person alive? Kid-ney transplants, yes. Heart transplants, yes, perhaps—but what if the patient is already old or has other ailments? When should a sick person be allowed to die? What *is* the "optimal lifespan"? Should the old and already sick be not the first but the last to receive COVID-19 vaccinations? We know that lifespan is subject to natural selection, so there must be an optimal—in the sense of most favorable to the continuation of the species—lifespan. What is it? What if it is shorter than the current average?

Politics are not immune from evolutionary optimality. What is the best political system? Most Americans assume that hierar-chy is bad, and the U.S. Constitution enshrines a limited democ-racy and the rights of the individual. However, the most stable (that is, evolutionarily successful) societies we know were not democratic and egalitarian but hierarchical and authoritarian. The ancient Egyptian culture survived substantially unchanged for thousands of years. The Greeks, the inventors of democracy, survived as a culture only for two or three centuries and were defeated by the undemocratic Romans, who lasted three or four times as long. The oldest extant democracy is less than three hundred years old. In the animal kingdom, the termites, ants, and bees, with built-in hierarchies, have outlasted countless more individualistic species.[13]

The attempt to base values on evolutionary success very soon raises questions about traditional beliefs.[14] The problem with "survival of the culture" as a value is that it requires reliable knowledge of the future. While some customs are clearly maladaptive under most imaginable circumstances, others are more contingent. The problem is that most of the prescriptions of traditional morality fall in the latter class. We simply do not know, belief by belief, custom by custom, rule by rule, whether or not our culture would, in the long run, be better off with or without them.

A similar argument has been made about business success in a capitalist economy. It may be that many businesses succeed not because of planning, not because they have foreseen a need and met it better or sooner than competitors, but simply because they have more or less accidentally adopted practices or products that, in the fullness and contingency of time, happen to work better than their competitors'.

It is certain that some cultures will survive longer than others. It seems very likely, moreover, that the ones that survive will have many beliefs that were in fact essential to their survival. But the importance of at least some of those beliefs *could not have been foreseen*, even in principle. The willingness of B. F. Skinner and E. O. Wilson to equate the *is* of future evolution to the *ought* of morality is an epistemological flaw, to which must be added the practical impossibility of evolutionary prediction. An evolutionary ethics is impossible for practical as well as epistemological reasons. And so, evolutionary imperialism is simply false.

Faith Returns

Furthermore (and this will not please your average atheistical social scientist), the argument that demolishes evolutionary ethics also provides a rational basis for *faith*—although not, I hasten to add, for any faith in particular. The reason is not that a faith is true scientifically. The reason is that for a society to function at all, rules seem to be necessary, even in cases like the examples I have given, in which certainty is (and perhaps must be) lacking. We deter smoking, outlaw some drugs, emancipate women, tolerate or even celebrate the non-reproductive, and preserve life at almost any cost, even though the evolutionary consequences of these decisions are unknown and probably unknowable. If rules must exist—even for situations in which science provides no clear basis for choosing them—then some other basis for choice is necessary. That basis is, by definition, a matter of faith.

Faith is to a culture as instinct is to an animal. Instincts are necessary for an animal to cope with situations where it cannot afford to "wait and see" how things will pan out. Neither an organism nor a culture can afford to rely on learning as a guide to every action, especially if consequences are long delayed or uncertain.

The evolutionary approach to the problem of values promises more than it can ever deliver. Harris, Dawkins, Skinner, Wilson, and most other scientific imperialists are confident of their own values and believe them to derive from science. Hence, they think the ethical problem easier than it is—which allows them to try to persuade us that values that are so obvious to them in fact flow from science. In fact, all are matters of faith.

● ● ●

The issue should have been settled by David Hume in 1740: the facts of science provide no basis for values. Yet, like some kind of recurrent meme, the idea that science is omnipotent and will sooner or later solve the problem of values seems to resurrect with every generation.

The science-based criterion most likely to achieve consensus, survival of the species or the culture, is impossible in practice because evolution is unpredictable. Embarrassingly, many practices that seem to favor survival are opposed to contemporary Western values.

Cultures depend on practices and beliefs. We do not know which of them in fact promote survival and which do not. We do know that without at least some of them, no culture can long survive. We must have faith in some unprovable things, but science cannot tell us what they should be.

CHAPTER 4

Was Darwin Wrong or Just Misunderstood?

E volution plays a central role in most debates about science. It's worth trying to see what it is and what it is not. How well do we understand it, really?

Christopher Booker (1937–2019) was a contrarian English journalist who wrote extensively on science-related issues. He produced an excellent critical review of the anthropogenic global warming (AGW) hypothesis and cast justifiable doubt on the allegedly lethal effects of low-level pollutants like airborne asbestos and second-hand tobacco smoke.[1] Booker has also lobbed a few hand grenades at Darwin's theory of evolution.

In a 2010 article, Booker covered a seminar of Darwin skeptics, many very distinguished in their own fields. These folk had faced a level of hostility from the scientific establishment which seemed to Booker excessive or at least unfair. Their discussion provided all the ingredients for a conspiracy novel:

They had come up against a wall of hostility from the scientific establishment. Even to raise such questions was just not permissible. One had been fired as editor of a major scientific journal because he dared publish a paper sceptical of Darwin's theory. Another, the leading expert on his subject, had only come lately to his dissenting view and had not yet worked out how to admit this to his fellow academics for fear that he too might lose his post.

The problem was raised at an earlier conference:

A number of expert scientists came together in America to share their conviction that, in light of the astonishing intricacies of construction revealed by molecular biology, *Darwin's gradualism* could not possibly account for them. So organizationally complex, for instance, are the structures of DNA and cell reproduction that they *could not conceivably have evolved just through minute, random variations.* Some other unknown factor must have been responsible for the appearance of these "irreducibly complex" micromechanisms, to which they gave the name "intelligent design" [emphasis added].[2]

I am a big fan of Darwin. And yet, "political correctness" surrounding Darwinism is a cause for concern. Skeptics should be able to state their objections without fear of reprisal. Their claims should be disputed, not demonized. That way, the contradictions they highlight can be resolved if we look more carefully at what we know now—and at what Darwin actually said.

The Logic of Evolution

The idea of evolution preceded Darwin. No serious critic of Darwin questions the fact that all organisms are related and that the living world has developed over many millions of years. Darwin's contribution was to propose, and support, a process—natural selection—by which evolution happens. It is the supposed inadequacy of this process that exercises Booker and other critics.

There are three parts to the theory:

1. Evolution itself: the fact that the human species shares common ancestors with the great apes. The fact that there is a phylogenetic "tree of life" which connects all species, beginning with one or a few ancestors who successively subdivided or became extinct in favor of a growing variety of descendants. Small divergences became large ones as one species gave rise to two, and so on.

2. Variation: the fact that individual organisms vary—have different phenotypes, different physical bodies, and different behaviors—and that some of these individual differences are caused by different genotypes, and so are passed on to descendants.

3. Selection: the fact that individual variants in a population will also vary in the number of fertile offspring to which they give rise. If number of offspring is correlated with some heritable characteristic—if particular genes are carried by a fitter phenotype—then the next generation may differ phenotypically from the preceding one.

Notice that in order for selection to work, at every stage the new variant must be more successful than the old.

Here are a couple of examples where natural selection can be seen to work. Rosemary and Peter Grant looked at birds on the Galapagos

Islands. They studied populations of finches and noticed surprisingly rapid increases in beak size from year to year. The cause was weather changes which affected the available food: mostly hard nuts in dry years, easier ones in wet. A prolonged drought for a few years favored birds with larger beaks better able to crack tough nuts. Natural selection operated amazingly quickly, leading to larger average beak size within just a few years.[3] Bernard Kettlewell had observed a similar change, over a slightly longer term, in the color of the peppered moth in England. As tree bark changed from light to dark to light again as industrial pollution waxed and waned over the decades, so did the camouflage color of the moths.[4] There are several other "natural experiments" that make this same point.

Looked at from one point of view, Darwin's theory is almost a tautology, like a theorem in mathematics:

1. Organisms vary (have different phenotypes).
2. Some of this variation is heritable, passed from one generation to the next (different genotypes).
3. Some heritable variations (phenotypes) are fitter (produce more offspring) than others because they are better adapted to their environment.
4. Ergo, each generation will be better adapted than the preceding one. Organisms will evolve.

Expressed in this way, Darwin's idea seems self-evidently true. But the simplicity may be only apparent.

The Direction of Evolution

Explaining nature by the interaction of two opposed forces was Darwin's style, long before he applied the method to his theory of natural selection. Even before he ever saw one, he thought up a theory of coral reefs that explained them by the slow sinking of the sea floor, countered by the slow upward growth of the corals.[5] He made a similar observation about the behavior of ants.[6] Darwinian evolution also depends on two forces: selection, the gradual improvement from generation to generation as better-adapted phenotypes are selected; and variation, the set of heritable characteristics that are offered up for selection in each generation. This joint process can be progressive or stabilizing, *depending on the pattern of variation.* Selection/variation does not necessarily produce progressive change. This should have been obvious, for a reason I describe now.

The usual assumption is that among the heritable variants in each generation will be some that fare better than average. If these are selected, then the average must improve, the species will change—adapt better—from one generation to the next. But what if variation only offers up individuals that fare the same as or *worse* than the modal individual? The *worse* will all be selected against and there will be no shift in the average; adaptation will remain as before. This is called *stabilizing* selection and is perhaps the usual pattern. Stabilizing selection is why many species in the geological record have remained unchanged for many hundreds of thousands, even millions, of years. Indeed, a forerunner of Darwin, the "father of geology" and Scot, James Hutton (1726–1797), came up with the idea of natural selection as an explanation for the *constancy* of species.[7] The difference—progress or stasis—depends not just on selection but on the range and type of *variation.*

The Structure of Variation

Many forget that Darwin's process has two parts, and that variation is just as important as selection. Indeed, without variation, there is nothing to select. But like many others, Richard Dawkins, a Darwinian fundamentalist, puts all weight on selection: "Natural selection is the force that drives evolution on," says Dawkins in one of his many television shows.[8] Variation represents "random mistakes," and the effect of selection is like "modelling clay."[9] Like Christopher Booker, he seems to believe that natural selection always works with small, random variations.

Critics of evolution find it hard to believe that the complexity of the living world can all be explained in this consistently incremental way. Darwin was very aware of the problem, as he writes in the *Origin of Species*: "If it could be demonstrated that any complex organ existed, which could not possibly have been formed by numerous, successive, slight modifications, my theory would absolutely break down."[10] But he was being either naïve or disingenuous here. He should surely have known that outside the realm of logic, proving a negative, proving that you *can't* do something is next to impossible.

Selection, natural or otherwise, is just a filter. It creates nothing. Variation proposes, selection disposes. All the creation is supplied by the processes of variation. If variation is not totally random or always small in extent, if it is creating complex structures, not just tiny variations in existing structures, then *it* is doing the work, not selection.

Looking at the biological phenomena Darwin sought to explain can help make sense of the distinction. Darwin was concerned about the evolution of the vertebrate eye: focusing lens, sensitive retina, and so on. The eye is an incredibly complex organ with a multitude of parts that all need to work together seamlessly for vision to take place. So how could the bits of an eye evolve and be useful before the whole perfect structure has evolved? Darwin pointed to the wide variety of primitive eyes in a range of species that lack many of the elements of the fully formed vertebrate eye but are nevertheless better than the structures that preceded them. These variations still have selective advantages, and then act as the basis for further variation.

There is general agreement that the focusing eye could have evolved in just the way that Darwin proposed,[11] although the underlying genetics remain a bit of a puzzle.[12] But there is some skepticism about many other extravagances of evolution: all that useless patterning and behavior associated with sexual reproduction in bower birds and birds of paradise, the unnecessary ornamentation of the male peacock, and many other examples of apparently maladaptive behavior associated with reproduction, even human super-intelligence—we seem to be much smarter than we needed to be as hunter-gatherers. The theory of sexual selection was developed to deal with cases like these, but it must be admitted that many details are still missing.[13]

Nonrandom Variation

In Darwin's day, nothing was known about genetics. He saw no easy pattern in variation, but was impressed by the power of selection, which was demonstrated in artificial selection of animals and crops. It was therefore reasonable and parsimonious—and consistent with the prevailing belief in *uniformitarianism*[14]—for him to assume as little structure in variation as possible. But he also discussed many cases where variation is neither small nor random. So-called "sporting" plants are examples of quite large changes from one generation to the next, "that is, of plants which have suddenly produced a single bud with a new and sometimes widely different character from that of the other buds on the same plant."[15] What Darwin called *correlated variation* is an example of linked, hence nonrandom, characteristics. He quotes another distinguished naturalist writing that "Breeders believe that long limbs are almost always accompanied by an elongated head," and "colour and constitutional peculiarities go together, of which many remarkable cases could be given among animals and plants."[16] Darwin's observation about correlated variation has been strikingly confirmed by a long-term Russian experiment with silver foxes selectively bred for their friendliness to humans.[17] After several generations, the now-friendly animals began to show many other features of domestic dogs, like floppy ears and wagging tails.

"Monster" fetuses and infants with characters much different from normal have been known for centuries. Most are mutants and they show large effects. But again, they are not random. It is well known that some inherited deformities, like extra fingers and limbs or two heads, are relatively common, but others—a partial finger or half a head, are rare to nonexistent.

The kinds of phenotypic (observed-form) variation that can occur depend on the way the genetic instructions in the fertilized egg are translated into the growing organism. Genetic errors (mutations) may be random, but the phenotypes to which they give rise are most certainly not. It is the phenotypes that are selected not the genes themselves. So selection operates on a pool of (phenotypic) variation that is not always "small and random."

Might "monsters" be the source of rapid evolutionary change: macro-mutation, as opposed to the more familiar micro-mutation? Probably not: most monsters die before or soon after birth. And a monster must find a mate if its characteristics are to survive. These difficulties sank the idea when it was first proposed many years ago, but more recent genetics research suggests that Richard Goldschmidt may have been on to something.[18]

Once in a very long while a nonrandom (that is, complex) variant may turn out to succeed better than the normal organism, perhaps lighting the fuse to a huge jump in evolution like the Cambrian explosion.[19] Stephen Jay Gould and Niles Eldredge rebranded George Gaylord Simpson's "tempo and mode in evolution"[20] as *punctuated equilibrium* to describe the sometimes-sudden shift from stasis to change in the history of species evolution.[21] Sometimes these jumps may result from a change in selection pressures. But some may be triggered by occasional large changes in phenotype with no change in the selection environment.

Another nonrandom option is *molecular drive:*

> An evolutionary process, like natural selection and neutral drift, that changes the genetic composition of a population, through the generations. It is distinct from natural selection and neutral drift . . .[22]

Geneticist Gabriel Dover in 1982 suggested that an "initial mutant sequence can increase or decrease in copy number within the lifetime of an individual," in other words covert changes in genetic potential can take place independently of selection. Dover proposes that the evolution of reproductively isolated species must operate in a much more subtle way than the simple "selfish gene" notion.[23]

Even mutations themselves do not occur at random. *Recurrent* mutations occur more frequently than others, and so would resist any attempt to select them out. There are sometimes links between mutations so that mutation *A* is more likely to be accompanied by mutation *B* ("hitchhiking"[24]) and so on. Genetic drift means that the constitution of a small gene pool may shift over time with no selection pressure at all.[25]

Is There Structure to Variation?

To convert Darwinism from a teleological to causal account, we need to understand an underlying mystery. How do species evolve when phenotypes are selected but genotypes are what is passed on? The mystery is: Just how is the information in the genes translated during development into the adult organism? Genes can have all manner of effects. The effect of a gene depends on its environment and other active genes; genes can control one another. A genotype is not a simple list of instructions; it is more like a computer program: an individual gene can participate in many different developmental processes.

How might one or two modest mutations sometimes result in large, structured changes in the phenotype? One possibility is suggested by recent interpretations of D'Arcy Wentworth Thompson's iconic 1917 book *On Growth and Form*.[26] Thompson "believed that evolution could sometimes advance in a leap rather than a shuffle," in the words of one commentator.[27] He noticed that the body shape of one species may be transformed into the sometimes-very-different shape of another by a simple geometrical transformation. Thompson's insights raise the possibility that small genetic changes may sometimes have large and coordinated effects.

Recent studies of the evolution of African lake fish suggest that there may be a predetermined range of possibilities, or at least limits, to the variety that can evolve from any given starting point. In Africa, genetically different cichlid fish in different lakes have evolved to look almost identical. "In other words, the "tape" of cichlid evolution has been run twice. And both times, the outcome has been much the same."[28] There is room, in other words, for the hypothesis that natural selection of small random variations is not the sole "driving force" in evolution. Some of the process, at least, may be guided by a pattern of variation that is not random but biased, directional.[29]

The laws of development (ontogenesis), if laws there be, still elude discovery. But the origin of species (phylogenesis) surely depends as much on them as on selection. Perhaps these largely unknown laws are what Darwin's critics mean by "intelligent design"? But if so, the term is deeply unfortunate because it implies that evolution is guided by intention, by an inscrutable agent, not by impersonal laws. As a hypothesis intelligent design is untestable, hence not scientific. (Do "mistakes" like the disease-prone vermiform appendix make the design unintelligent? Apparently not;[30] then what would?) Darwin's critics are right to see a problem with "small, random variation" Darwinism. But they are wrong to insert an intelligent agent as a solution and still claim they are doing science. Appealing to intelligent design just begs the question of how development actually works. It is not science, but faith.

● ● ●

Darwin's theory is not wrong. He knew, as many of his fans do not, that it is incomplete. Instead of paying attention to the gaps, and seeking to fill them, these enthusiasts have provided a straw man for opponents to attack. Emboldened by its imperfections they have proposed as an alternative "intelligent design": an untestable nonsolution that blocks further advance. Darwin was closer to the truth than his critics—and closer than some simple-minded supporters.

PART 2

The Profession

CHAPTER 5

Are We Losing Our Way?

S o why is science drifting from its Enlightenment moorings? Perhaps the answer lies in the structure of science as practiced today. Perhaps, counterintuitively, we are losing science by having too many scientists.

"The United States is producing more research scientists than academia can handle." So begins a July 2016 article by respected science reporter Gina Kolata.[1] It turns out that in recent years new Ph.D.s in science have a hard time getting a job similar to their mentor's: a tenure-track faculty position in a research university. Fifty years ago, in most areas of science, things were very different. Postgraduates, after four years of college, were able to get their Ph.D.s in four or five years. They usually got a tenure-track job at a reasonable university right after graduating.

Not now, though. An oversupply of nascent scientists has been the rule since at least 2010 and not just in the United States.[2] An *Economist* article titled "The Disposable Academic" described how "universities have discovered that PhD students are cheap, highly motivated and disposable labour."[3]

The norm now, in biomedicine and other science fields, is for newly minted Ph.D.s to take three or more one-year stints as post-doctoral fellows in other research labs before getting a tenure-track job. Depending on the discipline and their boss, they may have a chance to pursue some independent work without the distractions of teaching and administration that beset regular faculty. But it is more likely that they will serve simply as poorly paid help. In large, well-funded labs dealing in hot topics, postdocs and graduate students may be little more than over-specialized technicians. They no longer feel like students or academic colleagues. Collegial culture has been eroded. At the same time, colleges have exploited the plethora of under-employed Ph.D.s by hiring them as underpaid adjuncts. A predictable result is the rise of academic unions of the off-the-tenure-track disaffected.[4]

Yi Xue and statistician Richard Larson write that "in the STEM labor market: the academic sector is generally oversupplied."[5] Discussing the issue with Kolata, Larson commented that "84 percent of new Ph.D.s in biomedicine 'should be pursuing other opportunities.'" Top-flight engineering schools may get hundreds of applicants for a simple entry-level job, and Kolata concluded that even after years of marginally satisfying labor, "fewer than half" of science and engineering doctorates end up where they wanted to be in academe.[6] Some, especially in the "harder" sciences, do eventually get to use their talents in industry. But the rest have wasted years gaining skills that are largely irrelevant to their final profession.

The reason for the current oversupply is that the "reproductive rate" of academic scientists is very high. Grant-supported researchers will hire as many assistants, each a Ph.D.-in-waiting, as they can afford. Each lab director during his or her career produces five or more new scientists. As research support has grown, so have the number of Ph.D.s looking for a tenure-track job.

Perhaps this oversupply will eventually self-correct. Students considering graduate school may realize how unlikely it is that they will achieve an academic goal and choose accordingly. Should that occur, research grantees will begin to experience a labor shortage and the problem may eventually resolve itself.

Is the Frontier Really Endless?

But there is a problem that is potentially more serious and hard to pin down. In 1945 Vannevar Bush, engineer, public intellectual, and architect of today's system of government-sponsored academic research, wrote a famous report that led to the foundation of the National Science Foundation. In *Science, the Endless Frontier*, Bush declared:

> Scientific progress on a *broad front* results from the free play of free intellects, working on subjects of their own choice, in the manner dictated by their curiosity for exploration of the unknown [emphasis added].[7]

His unmistakable claim is that the field of science is essentially infinite, that the opportunities to make new discoveries are unlimited. In short: the more scientists the better!

But is that true? Bush's ambitious claim has come under attack recently,[8] partly because of problems with the research product. The ability to repeat an experiment and get the same results—*replication*—is the standard of truth in experimental science. And scientific theories are supposed to predict. There is a "replication crisis"[9] in social and biomedical science, and there are serious conceptual and predictive problems in economics.[10] Astonishing percentages of research results in areas from chemistry and biology to medicine and environmental science cannot be replicated. The figures in social science are probably even worse, but we do not know because exact replication is rarely attempted.

Experiments that cannot be replicated are not science but "noise" in the engineering sense. They just add confusion, proliferating rabbit holes into which new researchers are fruitlessly drawn.[11] Attempts are being made to remedy these problems, but their source may lie outside the control of the scientific community.

An obvious reason for these bad studies is the bad incentives under which most scientists operate. But ambition and the desire for fame are not necessarily obstacles to real science. James Watson, codiscoverer of the structure of DNA, was by his own admission competitive and ruthless in his quest.[12] Labeled by sociobiology icon E. O. Wilson as "the Caligula of biology" and "the most unpleasant human being" he had ever known,[13] Watson nevertheless succeeded because he tackled a real problem for which there was a real solution.

Too Few Soluble Problems?

The root cause of the replication crisis and "bad science" generally may lie beyond, and be a partial cause of, its present poor-incentive structure. Methods may be flawed because scientists in the current environment need to get results—something they can publish—replicable or not. But if the ratio of scientists to soluble problems is increasing, valid results may be harder and harder to get.

Vannevar Bush spoke of scientific advance on a "broad front." Broad, yes, but perhaps not infinite. As each problem is solved, new questions arise. There may be no end to this process, but the number of soluble problems at any given time is likely to be finite. "Knowledge" is not a product like cars or widgets: add more workers, make more widgets. Add more scientists, get more papers, maybe; but more knowledge? Maybe not.

The growing number of pseudo-scientific missteps we have witnessed in recent years may be not just a testament to human frailty, but a signal that the number of solvable scientific problems has not kept pace with the growing number of scientists. This disparity is not disastrous. There are still answers to be found; advance continues. But the mismatch does mean that the ratio of unsuccessful to successful experiments will increase.

A high rate of failure is not in itself a problem, scientifically speaking. Failure is okay; it is a necessary part of science. Almost every great scientific discovery was preceded by a long string of failures, all a testament to the honesty and thoroughness of the discoverer. But it is not just scientific discovery that is at stake: *repeated failure is not compatible with career advancement*, and science is now for most scientists a career, not a vocation. Failures are essential to scientific advance, but they do not play well with peer-review committees. An ambitious scientist cannot afford to fail.

And this reality has created a major problem, one that threatens to corrupt science to its core. Anxious researchers will be drawn to research methods that look enough like science to become accepted practice and are guaranteed to get positive results at least some of the time. Hence the reliance on the NHST method,[14] significance testing at a too-generous 5-percent level,[15] and the resulting replication crisis.

In other words, the replication crisis and other problems of science, like the apparent slowdown in the rate of discovery of new therapeutic drugs,[16] may reflect something more than human susceptibility to bad incentives. It is possible to do good science for less-than-noble motives. Perhaps the problem is not people, but *nature*. Perhaps there are simply too many scientists for the number of soluble problems available. Perhaps we have taken the low-hanging fruit and what is left is too tough to harvest without abandoning rigor.

There can be too much of anything. There must be an optimal number of scientists that is less than 100 percent of the adult population. Beyond that optimal number, the scientific community will begin to generate noise rather than signal, and advance is impeded. Are we at that point now in areas like social science and biomedicine? Vannevar Bush's inspiring prose was appropriate at the end of World War II and led to great advances in government-supported pure and applied science. But the situation now may be very different. We should at least be thinking about whether we need not more, but fewer, scientists.

Anti-Science *at the National Science Foundation*

This book is about science, not morals or politics. But in recent years a radical equity-first activist ideology has begun to pervade government science entities such as the National Science Foundation and the National Institutes of Health.[17] *Equity* is rarely defined, but in practice it seems to mean a close match between the proportion of a given race/identity in a given profession and its proportion in the general population. Since setting a target like this will affect the type and quality of future research, equity programs deserve scrutiny.

The National Science Foundation has a simple mission, summarized on its website as:

> To promote the progress of science; to advance the national health, prosperity, and welfare; to secure the national defense; and for other purposes.[18]

The phrase "for other purposes" gives the NSF some leeway, and *racial equity* has been a factor in the awarding and funding of research proposals for some years, to the point that academic virologists told Heather Mac Donald in 2020 that they "could receive hundreds of thousands more taxpayer dollars if they could find a 'diverse' student to add to the project."[19] Unproven assumptions about systemic racism, implicit bias, and the idea that race and gender diversity is equivalent to intellectual diversity have been so repeated in the bureaucratic community that they have become truisms, matters of faith. But, true or not, they should be irrelevant to an agency whose primary concern is the excellence of its researchers. A demonstrably creative community may be racially diverse, but not all racially diverse communities are creative. In short, ethnic and gender diversity is a poor proxy for scientific creativity.[20]

In fact, identity politics is booming at the National Science Foundation. In the summer of 2021, for example, proposals were invited for a program devoted to "Racial Equity in STEM Education." The synopsis reads in part:

> Persistent racial injustices and inequalities in the United States have led to renewed concern and interest in addressing systemic racism. The National Science Foundation (NSF) Directorate for Education and Human Resources (EHR) seeks to support bold, ground-breaking, and potentially transformative projects addressing systemic racism in STEM. . . . Core to this funding opportunity is that proposals are led by, or developed and led in authentic partnership with, individuals and communities most impacted by the inequities caused by systemic racism. *The voices, knowledge, and experiences of those who have been impacted by enduring racial inequities should be at the center of these*

> *proposals. . . .* Competitive proposals will be clear with
> respect to *how the work advances racial equity* [emphasis
> added] . . .[21]

Apart from the inflated language, there are scientific and policy,
not so say moral, problems with this statement. First, it is
activism—*racial equity* is to be advanced—not inquiry. Activism has
never been part of the National Science Foundation's mission. The
National Science Foundation exists to discover knowledge, not to
advance social justice. Second, proposers must be "individuals and
communities most impacted by the inequities caused by systemic
racism. . . . [their] voices . . . should be at the center of these proposals." In other words, principal investigators are to be qualified not by
their accomplishments but by their identities, which is the absolute
antithesis of the normal criterion by which National Science Foundation grantees are judged—and, arguably, racist to boot.

The phrase "systemic racism" occurs four times, and "systemic
barriers" once, all in the first 217 words of the program Synopsis. It
is never defined: millions of dollars of public money are to be
expended in eliminating an undefined something whose existence
is not questioned (never mind proved).

Here is another 2021 example, from the National Science Foundation Division of Human Resource Development:

> The U.S. National Science Foundation is investing in
> the establishment of five new NSF INCLUDES Alliances
> to enhance preparation, increase participation and
> *ensure the inclusion* of individuals from historically
> underrepresented groups in science, technology, engi-
> neering and mathematics education. This investment of
> $50 million is part of an NSF-wide effort to address

diversity, inclusion and participation challenges in STEM at a national scale [emphasis added].[22]

This is another activist (*ensure the inclusion*) project explicitly de-emphasizing merit in favor of racial and other identities. Once again, individuals are to be supported based on their membership in "underrepresented groups." For example, a recent award, totaling almost $10 million over five years, is for something called "The Alliance for Identity-Inclusive Computing Education (AIICE): A Collective Impact Approach to Broadening Participation in Computing." The project is proudly pro-minority—minority race and minority sexual identity. The abstract begins:

> While nationwide enrollment in undergraduate computing programs continues to increase, computer science (CS) is still overwhelmingly dominated by white and Asian, able-bodied, middle to upper class, cisgender men. Effects of this lack of diversity are evident in academic/workplace cultures and biased/harmful technologies (e.g., facial recognition, predictive policing, public services decisions, healthcare, and financial software) that negatively impact and exclude non-dominant identities. Despite this, identity (as defined in social science) is rarely, if ever, included in CS curricula, pedagogy, research, and policies.[23]

The project is justified in two ways, neither of them scientific. The desirability of reducing the proportion of "white and Asian, able-bodied, middle to upper class, cisgender men"—presumably in favor of transgender and homosexual men, the disabled, and people of color—is taken to be self-evident. Is there actual discrimination against these "non-dominant identities"? No evidence is provided or even hinted at. What is the proper proportion of individuals who are transgender, non-white, non-Asian, et cetera in computer science? Apparently it is too low, but how do we know? Merit is never mentioned. The aims of this proposal are unabashedly racist.

The other justification is that the dominant "white males," et cetera in the computing community are responsible for some bad developments, like programs for facial recognition and the prediction of crime. The proposal alleges a cause-effect relation: whites are somehow responsible—by virtue of, if not of their whiteness, of their not being in one of the groups preferred by the proposers—for the supposed evils of face recognition and predictive policing. No proof is offered that "whiteness" leads to evil results, or that its opposite to good. Indeed it is hard to see what such proof might look like.

More importantly, in addition to asserting causation for which there is absolutely no evidence, the justification is moral, not scientific. Who is to say that predictive programs are evil? Moral judgement is the province of politics and civil society, not a division of the National Science Foundation. It is conceivable that the National Science Foundation might set up a program, involving ethicists, religious scholars, legal experts, and politicians, as well as scientists—the kind of scheme used to evaluate the safety and efficacy of DNA modification, like the Asilomar conference[24]—to study such questions. In the absence of anything like that, moral claims are completely inappropriate for a scientific proposal. The proposal is not scientific in any sense at all.

Projects like this damage real science in several ways. First, activism has no place at an agency supposedly devoted to scientific inquiry. Activism is easy and science is tough; activists are always certain, scientists are often in a state of doubt. Once activists are allowed in, they will soon dominate. Second, in the search for "equity," scientific merit seems to be irrelevant. Hence grants like this will add to the science "community" people who are neither interested nor qualified in science. Scientific standards will therefore be sidelined, and the quality of research will suffer. Finally, $50 million (the EHR Equity Program) or more in taxpayer dollars that could have been spent on actual science are being wasted on endeavors that irrelevant and destructive, not to say unethical.[25]

• • •

Science now plays a larger role, and engages more people, than ever before. There may even be too many scientists in relation to the soluble problems available at any time. There certainly are too many aspiring scientists for the available front-line research positions. These are normal market disequilibrium problems. What is not normal is the increasing takeover of large government science agencies by a radical social-justice ideology. This movement threatens the quality of scientific research and diverts it from scientifically important problems to ideologically driven issues.

Science, in its modern form, is funded by a few large bureaucracies. Bureaucracies tend to grow, and they rarely become more lively and creative as they do so. When science was at its most productive, support came from many independent sources, not a monopsony, as it does now. This is the kind of diversity we should perhaps encourage.

CHAPTER 6

Scientific Publishing

Incoherent, Expensive, and Slow

C ommunication is essential to science. The aim of scientific publication is to convey new findings as quickly as possible to as many interested parties as possible. Some findings are more solid, more reliable, and more interesting than others; some are more relevant than others to particular research questions. The communication system should signal the area of published research and its validity and probable importance.

Above all, editors should filter out false findings. In engineering jargon, the system should be noise-free. All, or at least the great majority of, findings should be true. Partly as a legacy of the discredited postmodernist movement and partly because it is the nature of science that any finding is subject to revision, many people are now suspicious of the word "true."[1] In experimental research, at least, the term replicable (repeatable) has replaced it. In science there is no substitute for repeatable results, even though certainty will always elude it. Researchers should be able to trust what they read.

When the scientific community was small, communicating science was relatively easy. Submitting an article did not cost money beyond its preparation cost. There were just a handful of places to send your report. Volunteer (usually anonymous), expert reviewers, selected by an editor and probably known to the author, were pretty prompt and you got a decision: accept, accept with minor changes, or reject. After publication, access was via a paywall, although this was a not a problem for most researchers since most institutions had subscriptions to relevant journals.

The basic structure for established scientific journals remains the same. Submissions are reviewed by unpaid volunteer peers and publish-or-not-publish decisions are made by an editor. But there have been other changes. Many journals, most offering peer review, are now "open access:" the submitter pays but the reader does not. There are now many more journals available, and the cost issue is looming larger.

Access

Editing, formatting, and printing have a cost, which, in early days, was usually borne by scientific societies. The resulting journals, such as the *Proceedings of the Royal Society* or *Science* (published by the American Association for the Advancement of Science), were widely read and reasonably priced. But as the financial structure of science has changed, so have pricing policies. Subscriptions for individuals are usually still reasonable. *Science* (first published in 1880) now costs just over $100 a year; *Nature*, a United Kingdom–based journal of similar vintage (1869) published by a commercial publisher is similar: Macmillan offers an annual subscription for £64 (about $84).[2]

Institutional subscriptions are another matter. The best a librarian can do to get the various Royal Society journals is to buy a ten-journal "Excellence-in-Science" package for $29,370! Another example, the prestigious *Journal of Neuroscience*, published by the Society for Neuroscience, has five levels of institutional subscriptions ranging from $3260 to $5990 per year. Even "niche" journals are not cheap. In 2020 *Behavioural Processes*, a modest Elsevier journal aimed at behavioral biologists that I edited for many years, costs $4654 a year.

Again, cost was not a problem so long as the number of scientific journals was modest. Now it is not. Estimates vary, but the number of scientific journals worldwide is in the tens of thousands.[3] Wikipedia lists over one hundred in psychology alone,[4] and that doesn't include the new journals that pop up in my inbox almost every week. Giving its workers comprehensive journal access imposes an increasingly unaffordable financial burden on any research institution. The alternative, pay-to-publish in open-access (OA) journals, shifts the burden without reducing it.

The whole scientific publishing business is in fact a carryover. The present structure is a historical accident; scientific publishing would never be designed this way today. Much of the system is unnecessary. That it must change is certain. What is less certain is the exact nature of a viable alternative.

In the meantime, those who must pay—academic institutions and research grantors—are beginning to say "Enough!" In the United States since 2013, government granting agencies have required their grantees to place published work in an open-access site, a website that can be accessed for free by anyone. (The National Institutes of Health had made the same requirement in 2005.) But paywalled publishers were allowed a year after publication to do so. Since time and priority are vital in science, the one-year delay still gives the publisher a competitive edge.

A group of eleven European funding agencies has gone further. Their joint Open Access (OA) initiative ("Plan S") begins:

> After 1 January 2020 scientific publications on the results from research funded by public grants provided by national and European research councils and funding bodies, must be published in compliant Open Access Journals or on compliant Open Access Platforms.[5]

Ten "Principles of Plan S" follow.[6] There are eighteen other European science-support agencies that are not party to this agreement. Only one UK agency is listed (although it comprises seven research councils, so it probably embraces most UK science funding). No German science agency has yet signed up to the initiative.[7]

It is hard to judge how effective this initiative will be. Existing publishers object strongly, viewing the proposal as a threat to their business model.[8] The journal *Nature* announced to a UK Parliamentary Inquiry into open access as early as 2004 that "under a pay-to-publish system, *Nature* would have to charge authors between £10k and £30k per article,"[9] a ridiculous sum.

The current model is exploitative;[10] but it is also vulnerable.[11] Publishers exploit the unhealthy incentive structure of modern science. Scientists must publish, preferably in so-called "high-impact" journals like *Science* and *Nature*, but if not, *somewhere*. These journals are what economists call *positional goods*. That is, they are prestigious because they attract the "best" (most visible) papers. They attract the best papers because they are the most prestigious. In other words, elite journals exist in a self-reinforcing process which keeps those on top on top.

Lower-prestige journals can also succeed because the pressure to publish in a peer-reviewed journal—any peer-reviewed journal—is relentless and affects every active researcher. More evidence for the publish-or-perish culture of modern research institutions is the steady supply of new journals, many with relaxed standards for vetting submissions.[12] As one commentator put it, "Demand [to publish] is inelastic and competition non-existent, because different journals can't publish the same material."[13] And, of course, the material itself, the published article, costs the journal nothing. These features, a type of monopoly, plus zero-cost contributions and subsidized editing (those free peer reviewers), continue to make academic publishing very profitable, even though its vulnerability has been apparent for many years.[14]

The existing structure has yielded healthy profits for commercial publishers, but it is vulnerable for two reasons. First, the actual production of a published manuscript is now much cheaper than it used to be, because of word processing and online publication. Much copyediting is now automated. Authors can do most of what is necessary for publication by themselves, perhaps with a little formatting help from their institution.

Second, not only is there no need for hard copies and all those typesetters and printing presses, the majority of researchers also prefer *searchable electronic copies* of journal articles over paper ones. Paper scientific journals have been obsolete for some time.

What role remains for journal publishers? Well, the question worries the oligopolistic world of commercial science publishing,[15] which has been largely successful in blocking every major change to the system so far.

There are two reasons for the publishers' success. The first is simply tradition: once a prestigious journal, always a prestigious journal. Even as *Science* and *Nature* become increasingly expensive and develop their own political agendas, they remain a target for ambitious scientists. The other reason these journals and others like them persist is lack of a clear alternative. The sticking point is not the publishing process or access: the internet solves both of those. Almost every institution now offers its researchers an open-access site for their files. And finding relevant papers should also not be problem for readers. Google in some form or other can presumably locate papers by author, date, and keywords in a way that makes segregation by subdiscipline in journals unnecessary. The aggregation sites ResearchGate and Academia already perform this function and even notify users about relevant new posts. No doubt their algorithms could be improved and more control offered to users. The point is that the capability for intelligent search already exists. The problem isn't even editing: most institutions offer help in preparing research grants which could readily be extended to help with manuscript writing. (Not that there aren't other problems with helping people who should be capable of helping themselves!)

No, the problem is *vetting*. Researchers need to know not only about subject relevance and the general interest of a piece of work; they need some assurance as to its truth and probable importance. That is the job of peer review. The existing system is far from perfect. The increasing subdivision of social science has meant that the "peers" that are chosen to review papers are now likely to come from a small subgroup of the like-minded, so that papers that would not survive scrutiny by the scientific community at large have proliferated. Much of social science is now either obscure, false, or even nonsensical. Whatever the substitute for the present arrangement, we obviously don't want to encourage the segregation of subdisciplines that has led to this situation. Shielding manuscripts from review by qualified experts, or even knowledgeable and interested nonexperts, should not be part of any system.

Each of these suggestions raises questions. But one thing is certain: the present system is slow, expensive, and inadequate. Science needs something better.

CHAPTER 7

Peer Review and
"the Natural Selection of Bad Science"

The British journal *Nature*, home in 1953 to James Watson and Francis Crick's landmark DNA paper, was rather in the doldrums by 1966, with a backlog of submitted manuscripts and losing ground to the general-science leader, the U.S. journal *Science*.[1] That year, however, the publisher appointed as editor one John Maddox, a slightly eccentric theoretical physicist and science journalist. The review practices of the previous editor, Jack Brimble, were apparently erratic and informal, and at first Maddox continued in the same fashion. This was fortunate, at least for Watson and Crick, since under the more rigorous reviewing that Maddox began later "*Nature's* much more rigorous refereeing standards would mean that the [DNA discovery] paper as a merely speculative piece of work, would not have been accepted today!"[2] Maddox subsequently introduced a more conventional system of associate editors and ad hoc reviewers, possibly because of the sheer volume of work. *Nature* now competes with *Science* for the prize of most prestigious general-science journal.

There is little evidence that *Nature* was worse under solo management than it became under a more conventional regime.[3] The need for peer review in fact turns out to be relatively recent, and the admission that its earlier adoption would have led a journal to reject one of the most important papers of the twentieth century is cause for concern. It is perhaps required less for science than for the accountability needs of funding and career-controlling bureaucracies[4]—and the sheer number of scientists and submissions.

The now-standard system for accepting a manuscript submitted to a scientific journal—or awarding money to a supplicant researcher—is for the editor to get a review from independent experts. For each submission, the editor picks a couple of "relevant" experts and sends each a copy of the manuscript. These usually anonymous and always-unpaid reviewers send in comments and criticisms, necessarily after some delay, and recommend either acceptance, acceptance with changes, or rejection. When he has had time to check reviewers' comments, the editor makes a final decision. This is the famous peer-review process followed rather mechanically by reputable scientific journals (including, I am now slightly embarrassed to admit, the two I have edited).

In early days, science was a vocation for a small number of men who were of mostly independent means, like Charles Darwin, or who were employed in ways that did not seriously compete with their scientific interests. Isaac Newton was a professor at Trinity College, Cambridge, with light duties, for example. Evolutionary pioneer Alfred Russel Wallace had no private income, but he made a living collecting biological specimens in exotic locations, a profession that aided his biological work. Albert Einstein worked in the Swiss patent office, but the work was apparently not demanding, and he needed no funds to think. In the nineteenth and much of the twentieth century, the pace of science was leisurely and the number of scientists small. In those days, most journal submissions were reviewed just by the editor, unless he felt the paper was either controversial, questionable, or well beyond his expertise.[5]

The need for review was forced more by the cost of publication than by a flood of submissions. Transcribing, typesetting, printing, binding, and circulating hard copies cost money. There was a limit to the number of submissions that could be published. All this has changed in the twenty-first century. The number of scientists has vastly increased along with their competitive dependence on external sources of funds. Science has become a career rather than a vocation. Career scientists need to publish. The demand to publish is large and increasing every year.

On the other hand, since the advent of word processing, the internet, et cetera, the cost of publication has become negligible. So what is the problem? Why still such an emphasis on restriction and review? The obvious answer is that when there is huge competition for research positions and most researchers compete for funds from granting agencies, those in charge are on constant lookout for "objective" ways to rate applications and applicants. Since the time when science ceased to be a hobby or, more recently, a product of mission-oriented agencies like NASA and the military, science-support bureaucracies seek justification for their now-essential awards. Peer review has become the gold standard. But the new system has had many unintended effects.

Positive Feedbacks: "To Him That Hath Shall Be Given . . ."

There can be no more important goal for the research community in the next few years than to cut the link between publications and success . . .
–John Maddox, former editor of Nature[6]

One problem is that scientific publication is what economists call a *positional good*.[7] *Nature* and *Science* are top journals because the papers they publish are perceived to be exciting and high quality. The papers are good because these journals are perceived as the best and can therefore attract the best submissions, a positive-feedback loop. And "best" is quantified, post hoc, by something called the *impact factor*,[8] which rates a journal according to the number of *citations*—mentions in other published articles—that its articles receive, relative to the number (a number often negotiated with the journal publisher) of citable articles in the journal.

Impact factor is a flawed measure in many ways. Elite journals will be by definition the most read and their papers the most cited. Impact factor will be larger for papers in popular fields, larger in, say, neuroscience, than in ethology. It may not reflect the real contribution of a paper—a paper famous for its influential but fallacious results will have a high impact factor, as David Sarewitz notes:

> Consider, for example, a 2012 report in *Science* showing that an Alzheimer's drug called bexarotene would reduce beta-amyloid plaque in mouse brains. Efforts to reproduce that finding have since failed, as *Science* reported in February 2016. But in the meantime, the paper has been cited in about 500 other papers, many of which may have been cited multiple times in turn.[9]

The paper certainly had a high impact factor, but does that make it a good paper? Conversely, a paper that terminates a line of research by showing it to be based on a fallacy may be vitally important, but will attract few citations because scientists who might have worked in its area now look elsewhere.[10]

There is another positive-feedback loop in the social system of science. Scientific work costs money, much more now than in days past. Money comes in the form of research grants. Grants are awarded on the basis of competition among submitted research proposals. Competition is fierce. Researchers nowadays spend 50 percent or more of their time writing fifty-page proposals, often with a payoff probability as low as 10 percent. A proposal is more likely to be funded if its authors have a strong publication record in prestigious journals: publication yields money yields more publication—positive feedback again.

Pop-Up Journals

Getting published in some peer-reviewed journal is critical to career success in science. Demand is high, and, following an elementary axiom of economics, the supply of suitably crafted journals has expanded to meet it. A growing list of what I call "pop-up" journals has arisen to meet the need for publication. I get frequent email invitations to publish in, or even to edit, such a journal. Here is a very prestigious one I received at the end of 2020:

> Dear Dr. Staddon,
>   Bentham Science is an international publisher, providing academic researchers and industrial professionals with the latest information in diverse fields of education. Our peer-reviewed scholarly journals and books have an ever-increasing readership of millions of researchers worldwide.
>   The purpose of this letter is to bring an exciting project to your attention and seek your interest for it. Bentham

Science is planning to launch a new journal entitled "*Current Electrochemistry*"

Current Electrochemistry will publish original research articles, letters, reviews/mini-reviews, and guest edited thematic issues on all aspects of electrochemistry. The journal is essential reading for all researchers working in the field of electrochemistry.

In recognition of your outstanding reputation, we would like to nominate you for the position of **Editor-in-Chief** of the journal, and seek your consent. The responsibilities of the Editor-in-Chief are as follows [emphasis added] . . .

How these folk discovered my little-known expertise in electrochemistry is a mystery . . .[11]

This is not the first editor-in-chief offer I have received. Until this one, the most exciting was to be editor for the *Journal of Plant Physiology* (my emphasis). I study animals and people, not plants, but "tropism" pioneer Jacques Loeb discovered many "intelligent" characteristics of plants, so I was tempted to give the editorship a try.

The algorithmic promoters of these journals are like anglers, dangling a smorgasbord of tasty flies above schools of publish-or-perishers, hoping for enough bites to sustain at least some of their creations. This nonsense just shows how strong is the publication pressure on working scientists. The fact that these pop-ups work, that people pay to publish in them, and that these publications are *not routinely discounted* by promotion and grant-awarding committees, points to serious problems with the current culture of science. Predictably, these pressures have led to the occasional fraud[12] and the evolution of research methods that are more or less guaranteed to produce a publishable, if not a valid, result.

A 2017 article in the *Times Higher Education* points to two other problems with the present system, one soluble, the other not.[13] The easy one is that some papers don't show all the details necessary to repeat the experiment. This problem can be solved to some extent by vigilant reviewers. It can never be completely solved because occasionally the original experimenter may be unaware of a crucial detail. The great Greek philosopher Aristotle thought that heavy objects fall faster than light ones: dropping two objects of different weight but the same size and shape gives that result but leads to a wrong conclusion if you don't know about air resistance. Asking for more information—"What, exactly, did you drop?"—from an author is always legitimate.

Peer Review Problems

The tougher problem is that journal reviewers may reinforce a kind of scientific establishment. In the 2017 *Times Higher Education* article titled "Scientific Peer Review: An Ineffective and Unworthy Institution" the authors comment:

> Peer review is self-evidently useful in protecting established paradigms and disadvantaging challenges to entrenched scientific authority . . . [and] by controlling access to publication in the most prestigious [peer-reviewed] journals helps to maintain the clearly recognised hierarchies of journals, of researchers, and of universities and research institutes.[14]

Undoubtedly, advocates of an established paradigm have an edge in gaining access to a prestigious journal. Proponents of intelligent design will certainly encounter resistance if they try to publish in

Evolution, for example. If most scientists believe "*X*" to be true, it will take a lot to convince them of "not-*X*." This is inevitable, but perhaps the process has gone too far. If so, what are the alternatives? In the meantime science must live with some very bad side effects of the publication mania.

"The Natural Selection of Bad Science"[15]

Professor Brian Wansink was head of the Food and Brand Lab at Cornell University. The lab has had problems, some described in an article called "Spoiled Science" in the *Chronicle of Higher Education* early in 2017.[16] Four papers of which Wansink is a coauthor were found to contain statistical discrepancies. Not one or two, but roughly 150 discrepancies. That revelation led to further scrutiny of Wansink's work and to the discovery of other eyebrow-raising results, questionable research practices, and apparent recycling of data in at least a dozen other papers. All of this has put the usually ebullient researcher and his influential lab on the defensive.

More recently Wansink's lab published data purporting to come from eight-to-eleven-year-old children that was in fact obtained from three-to-five-year-olds.[17]

Compared to these gaffes the lab's next problem looks like a very minor one:

> Wansink and his fellow researchers had spent a month gathering information about the feelings and behavior of diners at an Italian buffet restaurant. Unfortunately, their results didn't support the original hypothesis. "This cost us a lot of time and our own money to collect," Wansink recalled telling a graduate student. *"There's got to be something here we can salvage"* [emphasis added].[18]

Four publications emerged from the "salvaged" buffet study. The topic is no doubt of interest to restaurateurs but unlikely to shed light on the nature of human feeding behavior. It's entertaining. The study is correlational not causal—no experiments were done. These are all characteristics typical of most of the "science" you will read about in the media: a distraction and a waste of resources, perhaps, but not too harmful.

The real problem, the probable source of all of Wansink's other problems, is hinted at by the bit in italics. It's pretty clear that Professor Wansink's aim is not the advancement of understanding, but the *production of publications.* By this measure, his research group is exceedingly successful: 178 peer-reviewed journal articles, 10 books, and 44 book chapters in 2014 alone. Pretty good for ten faculty, eleven postdocs and eight graduate students.

The drive to publish is not restricted to Professor Wansink. It is universal in academic science, especially among young researchers seeking promotion and research grants. The concept of the LPU ("Least Publishable Unit"means the least amount of data that will get you a publication so your total can be as large as possible. The analogy is to physical units such as BTU = "British Thermal Unit.") has been a joke among researchers for many years. (Even the number of authors per paper seems to have increased in recent decades, although I have not done an actual count.) I earlier described the new industry of "pop-up" journals that have arisen to meet this demand.

The emphasis in academic science on publishing is misplaced. Number of publications, even publications in elite journals, is not a reliable proxy for scientific accomplishment. Great scientists rarely have long publication lists; and a paper in an "elite" journal isn't necessarily a great paper. I will give just two examples. W. D. "Bill" Hamilton (1936–2000) was possibly the most important evolutionary biologist since Charles Darwin.[19] He published his first paper in 1963 and by 1983 had published a total of twenty-two, a rate of just over one paper a year. Several of these papers were ground-breaking. His discovery of the importance of what evolutionists call *inclusive fitness* was perhaps his most important contribution. But the number of papers he published is modest—compare them with Professor Wansink's prodigious output or Brian Nosek's promotion package described below. One paper a year would now be considered totally inadequate in most research institutions.

My second example is personal: my first publication, which was in *Science*.[20] The basic idea was that pigeons (the standard subject for operant-conditioning experiments) could follow the spacing of rewards: working hard for food when it came frequently, more slowly when it came less frequently. Here is what I found. The pigeon's output (response rate versus time) tracked the input cycle (reward rate versus time) beautifully. But, paradoxically, the pigeons worked harder when reward was infrequent (low points of the cycle) than when it was frequent (the high points). An older colleague pointed out a possible artifact, but I could find no evidence for his suggestion at the time.

It turned out he was in fact right; I confirmed his idea much later with a better recording technique. Pigeons do track rewards, but they track (with a lag) in terms of something called wait time, not in terms of response rate. By the time I found that out, this area of research was no longer fashionable enough for publication in *Science*.

So why did *Science* publish what was in fact a flawed article? I think there were three reasons: the data were beautiful, very orderly, and without any need for statistics. Second, feedback theory was then very much in fashion, and I was trying to apply it to behavior. And third, the results were counterintuitive, an appealing feature for journal editors wishing to appear on the cutting edge.

Do top journals such as *Nature* and *Science* really publish the best work? Are they a reliable guide to scientific quality? Or do they just favor fashion and a scientific establishment, as the two writers in *Times Higher Education* claim? Nobel prize winner Randy Schekman and the many authors whose work is described in a 2013 review article coauthored by German researcher Björn Brembs agree that fashion is a factor but point to more important problems.[21] First, painstaking follow-up work by many researchers has failed to show that elite (or what Schekman calls "luxury"), high-rank journals reliably publish more important work than less selective journals. Brembs, Katherine Button, and Marcus Munfanò write:

> In this review, we present the most recent and pertinent data on the consequences of our current scholarly communication system with respect to various measures of scientific quality. . . . These data corroborate previous hypotheses: *using journal rank as an assessment tool is bad scientific practice* [emphasis added].[22]

The top journals are in fierce competition. Newsworthiness and fashion are as important as rigor. As Schekman says, "These journals aggressively curate their brands, in ways more conducive to selling subscriptions than to stimulating the most important research. . . . Science must break the tyranny of the luxury journals. The result will be better research that better serves science and society."[23]

Acceptance criteria for elite journals do not, perhaps cannot, provide, a perfect measure of scientific excellence. Impact factor (journal rank) is an unreliable measure of scientific quality, for reasons I described earlier. Elite journals favor big, surprising results, even though these are less likely than average to be repeatable. They will also be cited more,[24] which has its own costs, as I described earlier. Neither where a scientist publishes (journal rank) nor how often he publishes (the length of his CV)—the standard yardsticks for promotion and the awarding of research grants—is a reliable measure of scientific productivity.

The present system has additional costs: the peer-review process takes time, and often several submissions and resubmissions may be necessary before an article can see the light of day. The powerful incentives for publication-at-any-price make for "natural selection of bad science," in the words of one commentator.[25]

Efforts to change the system are underway. Here is a quote from a thoughtful, if alarmingly titled, new book on the problems of science—*Rigor Mortis: How Sloppy Science Creates Worthless Cures, Crushes Hope, and Wastes Billions*:

> Take, for instance, the fact that universities rely far too heavily on the number of journal publications to judge scientists for promotion and tenure. Brian Nosek [who is trying to reform the system] said that when he went up for promotion to full professor at the University of Virginia, the administration told him to print out all his publications and deliver them in a stack. Being ten years into his career, he'd published about a hundred papers. "So my response was, what are you going to do? Weigh them?" He knew it was far too much effort for the review committee to read one hundred studies.[26]

Clearly change is needed. Science administrators can change right away by putting less emphasis on quantity and place of publication and giving much more attention to what aspiring researchers' papers actually say.

What Next?

How to vet submitted manuscripts, how to reform the inefficient, and in many ways corrupt, journal system, how best to limit costs—all are tricky questions. Here are some suggestions as starting points for debate:

- Consider abolishing the standard hard-copy journal structure.
- Suppose that all submissions, suitably tagged with interest-area labels by the author, were instead to be sent to a central repository. (A pirate site[27] containing "scraped" published papers, Sci-Hub,[28] already exists, and there are several open-access repositories where anyone can park files.[29])
- Suppose also that anyone who is interested in reviewing manuscripts, usually but not necessarily a working scientist, were to be invited to submit his (or her) areas of interest and qualifications to another repository.
- Reviewing would then consist of somehow matching up manuscripts with suitable reviewers. Exactly how this should be done would have to be determined. How many reviewers? What areas? How much weight should be given to matching reviewers' expertise, in general and in relation to the manuscript to be reviewed? What about conflict of interest, and so forth? How quickly should reviewers be expected to submit their review? (The present process is way too slow.) But if rules could be agreed on, the process could certainly be automated.
- Reviewers would be asked to both comment on the submission and give it two scores for: (a) Validity—are the

results true/replicable? (b) Importance—a purely sub-jective judgment.

- If a reviewer detects a remediable flaw, the manuscript author would have the opportunity to revise and resubmit and hope to get a higher score.
- Manuscripts would always publicly available, unless withdrawn by the author. But after review, they would be tagged with the reviewers' evaluation(s). No manuscripts need be "rejected."
- Interested readers could search the database of manuscripts by publication date, reviewers' scores, topics, et cetera in a more flexible and unbiased way than the current reliance on a handful of journals allows.

PART 3

Climate Change

CHAPTER 8

Is the Climate Warming?
Is There More Extreme Weather?

B ad science and the costs of our broken science system are most apparent in the debate surrounding climate change. Climate change has to some extent become a political platform masquerading as science.

This chapter and the next are an appropriate transition from the previous section, on biological science, to the next section, on social science. Climate science is, or should be, part of physical science.[1] But in recent years, as with almost any science that deals with important questions that cannot be decisively answered, passion has begun to win out over fact. Confronting apparently vital questions that well fit Weinberg's trans-science, climate science is both shaken and stirred by politics and social issues.

These chapters began with an article I coauthored with a colleague, Peter Morcombe, an electrical engineer.[2] He had long been skeptical, on purely scientific grounds, of alarmist carbon-dioxide-induced-warming claims. He persuaded me (very much a nonexpert, not much interested in the question but well aware of the heat generated by the issue,[3]) to look at the data, most of it freely available on the internet. We ended up with a case not against but mildly for carbon dioxide.

Soliciting comments, I sent the draft article to three climate scientists, two very well-known and the third a colleague at Duke's Nicholas School of the Environment. The first two were generally approving but did not comment in detail. The third, my Duke colleague, in a dismissive email, gave us no specific comments but basically questioned our right to address the issue at all. In effect, "stay in your lane" was his very unhelpful message.

Even quantum electrodynamics can be explained, at least in outline, to an educated if nonspecialist audience.[4] Climate change should be no different, especially as higher mathematics is not required and relevant data are widely available. My colleague's rejection of our attempt to understand the problem was anti-science and completely illegitimate. But it gives an idea of the nature of this subject, teetering on a knife's edge between knowledge and belief, science and faith.

Climate Alarm

Despite hints of a "pause" in recent decades,[5] there is general agreement that the earth's climate is warming, especially in the northern hemisphere. A tiny town in Greenland, Ilulissat, is seeing icebergs crumbling off the nearby Jakobshavn Isbrae Glacier more rapidly each year. The effects are not all bad, though. Tourism is booming in Ilulissat, access to metal-ore sites is easier, and the fishing harbor can be used for most of the year.[6]

The consequences elsewhere are less benign, however. A warming climate causes sea level to rise by thermal expansion and by the melting of land ice in Greenland and elsewhere. Low-lying settlements can be flooded. On the other hand, the Antarctic, containing as it does some 90 percent of the planet's ice, is the biggest threat and shows less sign of warming, except at its edges.[7] Recent reports even suggest that Antarctic sea ice is growing.[8] Nevertheless, if sea level rises, many coastal communities will be under stress. (Just to confuse things, the latest reports seem to show the Jakobshavn Glacier growing again,[9] but Peruvian glaciers continue to melt.[10])

Despite these uncertainties, the consensus is that the planet is warming and will continue to do so, that this warming will cause the sea level to rise and will also increase the frequency of severe weather events like hurricanes, tornadoes, and wildfires. Many people, from a troubled Swedish teenager to the president of Harvard University, not to mention the head of Alphabet[11] and U.S. President Joe Biden,[12] say we are facing a crisis of biblical proportions:

> We must face up to the stark reality of climate change. The scientific consensus is by now clear. . . . Climate change poses an immediate and concrete test of whether we . . . will fulfill a sacred obligation: to enable future generations to enjoy, as we are privileged to enjoy, the wonders of life on Earth.[13]

"Listen to the scientists . . . unite behind the science . . . .take real action," an angry Greta Thunberg told U.S. lawmakers in 2019.[14]

I agree that any political action on climate should begin with science. The problem is that "the science" is often equivocal. There are passionate advocates on both sides of the climate issue. There are websites (of so-called "climate deniers") skeptical of any human agency in climate change,[15] if not of climate change itself;[16] there is at least one website solely devoted to criticism of skeptics,[17] as well as many presenting, or simply assuming, the case for human-caused change.

There is in fact no scientific unanimity about the causes of climate change or even about the magnitude of the change.[18] Understand also that *consensus is irrelevant*. Science is not a democracy; truth cannot be decided by vote. That a majority of scientists agree may be important to a politician who has to take action on an issue. It should be irrelevant to an inquirer who seeks the truth. The validity of a conclusion depends on the relevant facts and the arguments they support, not the number of people cheering each side. There are many examples in the history of science—from phlogiston theory to continental drift, not to mention confident contemporary prescriptions about a healthy diet[19]—where the consensus was in fact wrong.

These chapters are an effort to present an objective, reasonably comprehensive summary of the science, bearing in mind the multiplicity of sources and the fact that passionate politics affects almost all of them. Our reading has brought us to conclusions that are far from apocalyptic. There are even reasons for optimism. Is "aggressive action"[20] really required? We see nothing in the science that compels drastic measures. We may be wrong, but we have tried in as brief a

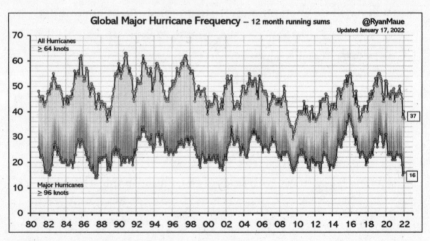

FIGURE 1. "Global Hurricane Frequency (all & major)—12-month running sums. The top time series is the number of global tropical cyclones that reached at least hurricane-force (maximum lifetime wind speed exceeds 64-knots). The bottom time series is the number of global tropical cyclones that reached major hurricane strength (96-knots+)." Source: Ryan Maue, "Global Major Hurricane Frequency," February 21, 2022, http://climatlas.com/tropical/.

space as possible to explain why we are skeptical of a fictional consensus. Let's look first at trends.

Hurricanes, Tornadoes, and Droughts: Have They Increased?

Perhaps because images and videos of dramatic events are now so widely publicized by the media, and perhaps because so many people are primed to see climate change as a cause of any weather disaster, many think that the frequency of natural disasters has greatly increased in recent decades.[21] Do the statistics, the data, (and by data I mean long-term trends that adequately represent the planet as a whole, not just a particular region) actually support the popular impression?[22] A Google search reveals a dozen or more graphs of hurricane frequency over the last fifty or a hundred years. Figure 1 is one of the more detailed, just covering recent decades. It shows data on all hurricanes as well as just major ones—from 1980 to 2021. Hurricane intensity is derived by combining force and duration. Major hurricanes have increased a little, hurricanes as a whole have declined in frequency. There is no obvious trend.

Figure 2, from the U.S. Environmental Protection Agency (EPA), goes a little further back, to 1860, covering hurricanes and other tropical storms in the Atlantic Ocean, the Caribbean, and the Gulf of Mexico. Just the number of hurricanes is counted since the early data allows nothing more precise. But again, there seems to be no trend. The search shows many confirming, and few discrepant, graphs.[23] Most data agree that there is no trend over the past 150 years or so.

The political problem is not the hurricanes, but growth of population and structures in risky areas. A 2018 analysis summarized the data:

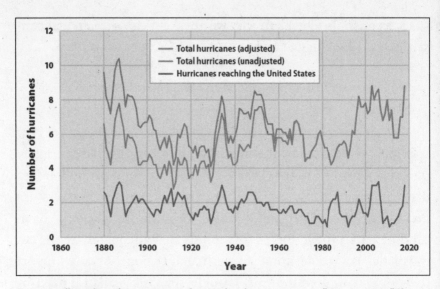

FIGURE 2. "Number of Hurricanes in the North Atlantic, 1878–2020." Source: EPA, "Climate Change Indicators: Tropical Cyclone Activity," https://www.epa.gov/climate-indicators/climate-change-indicators-tropical-cyclone-activity.

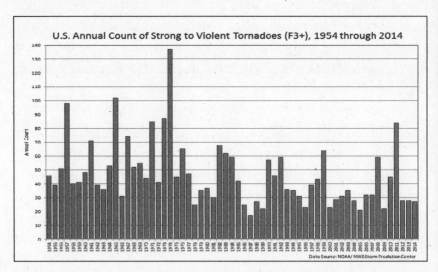

FIGURE 3. "Annual Strong to Violent Tornadoes (F3+) in the US, 1954 to 2018." Source: NOAA, in Mark J. Perry, "Inconvenient Weather Fact," April 22, 2019, https://www.aei.org/carpe-diem/inconvenient-weather-fact-for-earth-day-the-frequency-of-violent-tornadoes-fell-to-a-record-low-in-2018/.

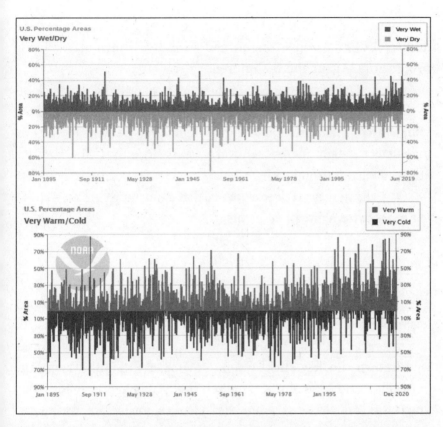

FIGURE 4. *Top:* U.S. data on very wet or dry years, from 1895 to 2019. *Bottom:* Comparable data for very hot or cold years. Source: NOAA, "U.S. Percentage Areas (Very Warm/Cold, Very Wet/Dry)," 2019–2020, https://www.ncdc.noaa.gov/temp-and-precip/uspa/wet-dry/0?submitted=View; https://www.ncdc.noaa.gov/temp-and-precip/uspa/warm-cold/0.

While neither U.S. landfalling hurricane frequency nor intensity shows a significant trend since 1900, growth in coastal population and wealth have led to increasing hurricane-related damage along the U.S. coastline.[24]

It seems to be the amount of damage, rather than the frequency of hurricanes, that has led to the public impression that hurricanes are becoming more frequent and more severe.

Figure 3 shows National Oceanic and Atmospheric Administration (NOAA) data on strong tornadoes in the United States, from
1954 to 2014. Again, there is no trend.

Figure 4 (*top*) is NOAA data showing very wet and dry periods
in the United States from 1895 to 2019, with no trend. Figure 4 (*bottom*) shows similar data for very warm/cold years. A sympathetic eye
may detect a slight increase in warm years since 1995. Similar data
are available for droughts.

Overall, there is no clear evidence that a warming climate is caus-
ing more extreme weather events.[25]

Temperature

There is some evidence that northern-hemisphere temperature
has increased in the past several decades (although the increase is
not unprecedented[26]). Figure 5 shows a simplified version of the
well-known "hockey-stick" graph of Michael E. Mann, Raymond
S. Bradley, and Malcolm K. Hughes, first published in 1999.[27] The
graph shows deviations ("anomalies") from a northern-hemispheric
mean temperature from AD 1000 through AD 2000. The older
temperatures had to be estimated by a complicated statistical model
using proxies, such as tree rings from a few specimens of a long-lived
Rocky Mountain pine, varved (layered) sediments, ice cores, corals,
and the like. Only the recent data are direct measurements. (Even
direct measurements are not without problems as indicators of global
temperature: completeness of coverage—the location of the ther-
mometers—"heat-island" effects of cities, seasonal variation, and
the like.[28])

The hockey stick caught on because it shows (ragged line) a
slow decline in hemispheric temperature from AD 1000, followed
by a dramatic increase. Estimated carbon dioxide (CO_2) levels

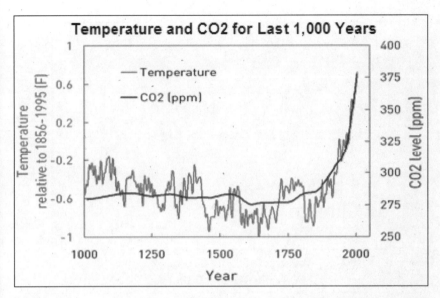

FIGURE 5. The northern hockey-stick graph of Mann et al. (1999) showing deviations from an average temperature (temperature anomaly and CO_2 levels from AD 1000 to AD 2000). Source: Bill Chameides, "Part 3 of 5: Causes of Past Climate Change," June 29, 2007, http:// blogs.edf.org/climate411/2007/06/29/human_cause-3/.
http://www.realclimate.org/index.php/archives/2004/12/temperaturevariations-in-past-centuries-and-the-so-called-hockey-stick/

(smooth line) follow the same trajectory, coinciding with the rise of Western industry after 1850 or so (more on this correlation in a moment).

The hockey-stick graph has been vigorously criticized.[29] After several disputes, including a long-running lawsuit,[30] a 2006 Congressionally sponsored report concluded that "*global mean surface* temperature was higher during the last few decades of the 20th century than during any comparable period during the preceding four centuries," even though the hockey stick represents data only from the northern hemisphere (emphasis added).[31] Unsurprisingly, there is still dissent.[32] For example, there is good historical evidence for a European medieval warm period (MWP) from about AD 950 to AD 1250, which was followed by a brief "little ice age."[33] This bump does not show up

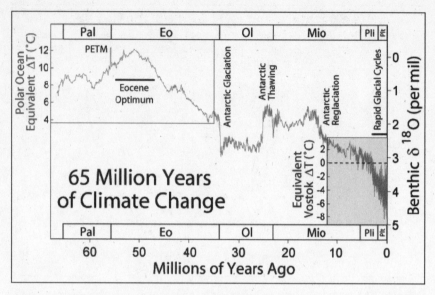

FIGURE 6. Global temperature during the last 65 million years, estimated from oxygen isotope composition of benthic foraminifera. Source: Wikipedia, "Paleocene—Eocene Thermal Maximum," January 1, 2022, https://en.wikipedia.org/wiki/Paleocene%E2%80%93Eocene_Thermal_Maximum.

in the hockey stick. Like the hockey stick, it gives information only about the northern hemisphere.

It does appear that the temperature of at least the northern hemisphere may have increased substantially, if not smoothly, over two periods during the last one thousand years: the MWP and the recent hockey-stick upturn. The MWP was followed by a temperature decline, but majority opinion expects the current increase to continue. I get to the reasons for this belief in a moment.

Figure 6 shows global temperature from 65 million years ago to the present.[34] Temperature is estimated by a proxy, but never mind the absolute scale, the graph clearly shows that temperatures in recent times, the last million years or so, were in fact much lower than they were in earlier periods. Hence, it seems likely that humanity, if not the present form of civilization, could endure a considerable rise in global temperature.

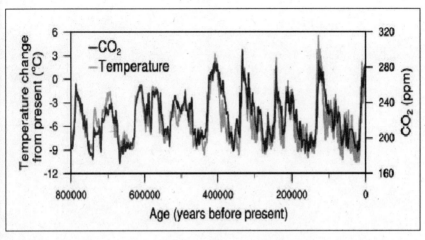

FIGURE 7. Temperature vs. CO_2—800,000 years ago until present. Dark gray line: atmospheric CO_2. Light gray line: temperature anomaly—"measured from the EPICA Dome C ice core in Antarctica." 2020 CO_2 level is 409 ppm, well above the terminus of this graph. Source: NOAA, "Temperature Change and Carbon Dioxide Change," October 2021, https://www.ncdc.noaa.gov/global-warming/temperature-change.

Figure 7 shows the covariation of atmospheric carbon dioxide and temperature, a topic we will return to in the next chapter. For now, just look at the light gray line, which shows temperature change (measured as it must be by a proxy method—Antarctic ice cores) relative to the present. Temperature has varied in quite large cycles from 800,000 years ago until now. Present temperature is close to a peak. Nevertheless, it is clear that the earth's temperature has been as high or higher than the present temperature during three or four earlier periods during the last 500,000 years and that previous peaks occurred in the presence of *Homo sapiens*, but without the aid of man-made CO_2.[35] Unless the recent temperature rise is not part of the historical pattern, but due to a unique new factor which may continue to force a rise, humans and higher temperatures can coexist—no apocalypse is in prospect.

CHAPTER 9

Carbon Dioxide and the AGW Hypothesis

A striking feature of Figure 7 in the previous chapter is how closely CO_2 concentration and temperature track one another. Is the CO_2 driving temperature or vice versa? The National Oceanic and Atmospheric Administration comments:

> A small part of the [CO_2 versus temperature] correspondence is due to the relationship between temperature and the solubility of carbon dioxide in the surface ocean, but the majority of the correspondence is consistent with a feedback between carbon dioxide and climate.[1]

In other words: the CO_2 ↔ *warming* causation goes both ways: an increase in CO_2, a *greenhouse gas*, causes atmospheric warming; atmospheric warming heats the oceans and releases dissolved CO_2: together these two processes constitute a self-reinforcing positive-feedback loop. The National Oceanic and Atmospheric Administration reflects the consensus that the ocean warming → CO_2 link is smaller than the CO_2 → warming link. On the other hand, if, contrary to the prevailing view, the warming → CO_2 link is the stronger, we need be less concerned about human additions to atmospheric CO_2.

There is of course another feedback cycle, involving water vapor.[2] Atmospheric warming evaporates more water from the oceans, and the increased water vapor traps solar heat even more effectively than CO_2. But water vapor also condenses into clouds, which reflect radiation and so have a cooling effect. And, unlike the total atmospheric CO_2, which is slowly increasing as a result of fossil-fuel combustion, there is no secular increase in total water vapor. Water is involved in complex feedbacks, but cannot be blamed for any temperature trend.

An obvious theoretical poser is that since both of the CO_2 feedbacks are positive, there should be no limit to temperature increase. We haven't fried yet, so there must be limiting factors:

(a) Temperature will equilibrate when it rises to the point that as much heat is lost by radiation into space as is absorbed from the sun. (b) Biological factors—plant and algal growth are a negative feedback (see seasonal variation in Figure 8 and leaf growth in figure 10): as CO_2 level rises, plants absorb it at a faster rate. (c) As land ice melts, planetary liquid water will increase, as will its capacity to absorb CO_2, a negative feedback. (d) Formation of carbonates, due to weathering and other processes, reduces CO_2 levels, another negative feedback. (e) Complex interactions with water vapor, methane, and other static greenhouse gases may have equilibrating effects. If the equilibrium temperature controlled by all these processes is livable, no worries.

On the other hand, if temperature is mostly driven by CO_2 (rather than the other way round) and CO_2 concentration has increased substantially since the advent of the industrial revolution 250 or so years ago, the conclusion seems both alarming and obvious: human-produced carbon dioxide, CO_2, is the cause of the global warming that has occurred over the past 150 years or so (Figure 5 in the previous chapter) and will likely cause more warming as CO_2 concentration continues to increase. This is the consensus *anthropogenic global warming* (AGW) hypothesis.

AGW is one possibility. The other is that global (or at least northern-hemisphere) temperature is just following the natural temperature cycle, like the peaks in Figure 7, in which case the rise in CO_2 is mostly caused by the rise in temperature, rather than the opposite. Human activity becomes irrelevant. The investor's motto "Past performance is no guarantee of future results" is appropriate. If the cause is not human, we must simply adapt to climate change since we cannot directly affect it. On the bright side, CO_2 levels may continue to rise without arousing undue anxiety.

The Keeling Curve[3]

AGW is the idea that has captured climate scientists' imaginations, partly because the putative cause has been precisely measured. The *Keeling curve*. Figure 8 shows atmospheric CO_2 levels measured at the Mauna Loa Observatory in Hawaii from 1958 to 2012.[4] The jiggles are seasonal variation: there is less CO_2 after northern summers because most plant life on the planet is in the northern hemisphere and plants consume CO_2. The inset shows the Keeling curve as usually displayed, with a truncated y-axis, which steepens the slope.

The curve shows a slow upward trend: atmospheric CO_2 levels have indeed increased in recent decades. Assuming that most of the correlation in Figure 7 reflects $CO_2 \rightarrow$ warming—*forcing*—rather than some other cause, the data are therefore consistent with the hypothesis that the recent (since 1900 or so) hockey-stick rise in hemispheric temperature is caused by the parallel rise in atmospheric CO_2. The assumption that $CO_2 \rightarrow$ warming is the major cause of the CO_2-temperature correlation is obviously the key supporting assumption for the AGW hypothesis.

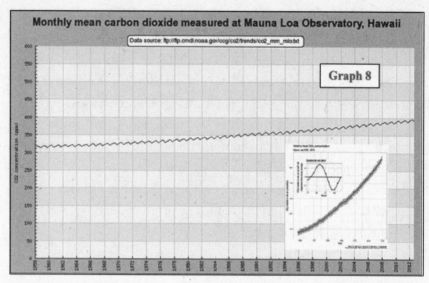

FIGURE 8. The Keeling Curve: CO_2 concentration from 1956 to 2012 at two scales—absolute or (inset) with truncated y-axis; cycle is seasonal variation. Sources: "Monthly Mean Carbon Dioxide Measured at Mauna Loa Observatory, Hawaii," JunkScience, September 6, 2012, http://junksciencearchive.com/MSU_Temps/MaunaLoaCO2.png; Wikipedia, "Atmospheric carbon dioxide (CO_2) concentrations from 1958 to 2020," https://en.wikipedia.org/wiki/Keeling_Curve#/media/File:Mauna_Loa_CO2_monthly_mean_concentration.svg

But is it true that recent temperature rise is caused by the rise in CO_2? CO_2 is less than 400 parts per million, 0.04 percent, of the atmosphere; it is slowly increasing, at least so far (Figure 8). Only about 4 percent of that 0.04 percent is caused by human activity.[5] CO_2 is much less effective as a greenhouse gas than water vapor (average 0–4 percent of the atmosphere), not to mention clouds.[6] But CO_2 is the focus: does the tiny human addition to the small amount of CO_2 naturally in the atmosphere really have a significant effect on the temperature of the planet, and how do we know?[7] And is it true that the predicted increase in the earth's temperature threatens human life? (Probably not, if the temperature data in Figure 6 in the previous chapter are to be believed.)

The earth would be uninhabitable but for an atmosphere that contains greenhouse gases, like water vapor and CO_2, which limit the amount of the sun's heat re-radiated back into space: "Without greenhouse gases, the average temperature of Earth's surface would be about -18° C (0° F), rather than the present average of 15° C (59° F)."[8] Some models suggest an even lower average. Greenhouse gases are essential to life, but have we now got too much of a good thing?

Models

There are two ways to find out. The most widely discussed way is via climate models. Climate modeling is obviously complicated; the outcome depends on the assumptions on which the model is based and the relative weights given to different factors. The effects of clouds, for example, are complex and omitted in some models. All this is quite well explained on several web sites.[9]

Scientists are not interested in detailed ups and downs of global temperature, but in trends: Is temperature increasing, and if so, by how much each year? The question is simple; the models are not. Most climate models and weather models are similar in that both are tuned to match existing data.[10] This is called retrodiction or hindcasting; an older term is simply *fitting*. The parameters of the model are adjusted until its output matches an existing data set as closely as possible. The tuned model is then used to predict. The models are not parsimonious in the usual sense of having few free parameters (also known as "fudge factors"). For example, one article comments: "The number of parameters N to be considered can easily be 10–30, although typically only a subset of these will have strong sensitivity."[11] In most scientific work with testable models, no more than three or four free parameters are acceptable, especially if the data to be predicted is just a monotonic trend and not, say, temperature movements decade by decade or even year by year. There is convincing criticism of widely used statistical assumptions.[12]

Weather and Hurricane Models

Any theory that deserves to be called scientific must make *testable* predictions. Climate models are often compared to weather models, but the differences are instructive. Weather models are tested every day; they are pretty accurate in predicting the weather tomorrow and perhaps a day or two after. Hurricane-track models are an intermediate case: they can only be tested when hurricanes occur, a few times a year. There are several hurricane models, and, as we might expect, they don't agree and are often very wrong,[13] although they have improved over the years.[14]

Testing a model is essential to improving its accuracy. Climate models are the worst possible case: they make predictions over decades or even centuries. From a practical point of view, *they are untestable*. How accurate can we expect them to be? Polymath theoretical physicist Freeman Dyson commented several years ago:

> Syukuro Manabe, right here in Princeton, was the first person who did climate models with enhanced carbon dioxide and they were excellent models. And he used to say very firmly that these models are very good tools for understanding climate, but they are not good tools for predicting climate. I think that's absolutely right.[15]

The difficulty of climate modeling is underlined by the number of different models that are available. There is only one model of planetary attraction, Newton's, and even it doesn't always yield an easy answer.[16] Climate models yield many answers, and we don't know which (or whether any) is correct. Nor does it make sense to average them, since we have no idea of the error distribution.[17]

We can illustrate the problems with retrodiction with a famous and for-a-while-successful model in a different area, the Black-Scholes financial model (BSM).[18] The dataset to which the model was fitted was past stock prices. The model gives a precise value for an option and did very well for a while, lifting the investment firm Long Term Capital Management (LTCM) into the financial stratosphere in just three years.[19] The Black-Scholes model is a good illustration of the retrodictive technique because all the parameters of the model but one can be directly measured. The one parameter that must be estimated from past data is *volatility*, the moment-to-moment variability in a stock price.

The Black-Scholes model worked well for Long Term Capital Management for three years, but then market volatility increased, for reasons outside the scope of the model. The model's predictions were wrong and the firm Long Term Capital Management collapsed, almost bringing the financial system down with it. Past performance really is no guarantee of future results. In other words, the parameters of a model that are estimated from past data may or may not be stable. The parameters in Einstein's equation, $E = mc^2$, m and c, mass and the velocity of light, are directly measurable and, under the appropriate conditions, invariant. But what are called *fitted parameters* are just estimates from past fits. They may change in the future, in which case model predictions based on them will be wrong.

Climate models are much more complicated than the Black-Scholes model. General circulation models (GCM), for example, divide the atmosphere and part of the ocean into millions of boxes (*voxels*); each voxel is set to certain initial conditions of temperature, pressure, wind, and atmospheric composition. These values are fed into a set of equations, making up millions of lines of computer code in the largest models. The equations change the values of the relevant dependent variables, such as temperature, pressure, and CO_2 concentration, from time-step to time-step, according to the processes they embody. The equations include well-established physical laws, like the Navier-Stokes equations, which describe the movement of viscous fluids, and the Arrhenius model for CO_2 temperature effects. Svante Arrhenius's 1906 estimate, based on laboratory studies and the equations of physical chemistry, was that doubling the amount of atmospheric CO_2 should increase atmospheric temperature by 4°C (although he revised it downwards later), which is close to what AGW advocates now believe.[20]

Arrhenius's estimate was based on a necessarily simplified system.[21] Many other variables are at play in the real atmosphere, of course. But no matter how elaborate the model, no matter how much past data it incorporates, the same restriction applies to all models with fitted parameters: the future may not be like the past.

Correlations

There is a second way to assess the temperature effects of CO_2: to look at the historical record and see if temperature and CO_2 concentration covary. They do at certain time scales, but not at others. Figure 7 shows close covariation over the past 800,000 years. But, as we are frequently reminded and almost as frequently forget, correlation is not causation. The covariation could mean that CO_2 causes warming, or that oceans, warming for some other reason, release more CO_2. The latter is a simple fact that is easily demonstrated and well understood: warm water can retain less dissolved gas than cool water. We also know that CO_2, tiny as its concentration is in the atmosphere, can reduce heat re-radiation into space—but by exactly how much is uncertain. Does the correlation in Figure 7 reflect heating aided by CO_2 increase? Or does the CO_2 just add a little bit to heating from another source?[22]

In fact, as I pointed out earlier, both effects seem to be operating. Global heating, caused by such things as small changes in the earth's orbit and in solar output, causes a rise in global CO_2, as it is "out-gased" from warming oceans. The evidence for this is an acknowledged 800-year lag: going back some 800,000 years, the temperature rise often *precedes* the subsequent rise in CO_2.[23] Sometimes warming lags CO_2 increase, sometimes the opposite.

> Comparison with available temperature records suggest that although CO_2 may have been a main driver of temperature and primary production at kyr or smaller scales, it was a long-term consequence of the climate-biological system, being decoupled or even showing inverse trends with temperature, at Myr scales.[24]

Abel Barral and others show data indicating that the correlation between CO_2 and temperature seems to break down at a long time scale. Figure 9 supports this idea.[25] The graph shows estimates of CO_2 concentration and temperature from the most remote past to the present. Bear in mind that the older the estimate, the less reliable it is likely to be. If CO_2 acts as a multiplier that can amplify small temperature changes into large ones, then we should see covariation between CO_2 and temperature, as in Figure 7. Figure 9 covers much larger time and temperature scales and smooths over the relatively short-term changes in Figure 7. Figure 9 shows that current levels of CO_2 are much lower than they have been in the past. There is no correlation—not even a negative correlation—between CO_2 level and temperature over much of the time period in this graph.

Taken together with the data of Barral and others, the data in Figure 9 provide no basis for anything more than a weak positive-feedback effect of CO_2 on global temperature. Can this feedback induce substantial warming by itself in the absence of exogenous forcing? Many, perhaps most, climate scientists believe it can. The data are not

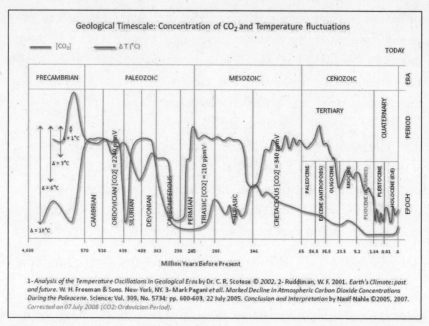

FIGURE 9. Temperature and CO_2 changes over a very long time period. Source: Hemlata Pant et al., "Geological Timescale," (July 2020), https://www.researchgate.net/publication/342589030_Innovations_in_Agriculture_Environment_and_Health_Research_for_Ecological_Restoration/figures?lo=1.

conclusive. The evidence that increases in human-produced CO_2 will force substantial increases in global temperature is in fact remarkably weak. The "AGW consensus" seems to rest largely on the intuitions of climate scientists. The data provide good support for climate change, but very weak support for a human-caused apocalypse.

How Bad Can It Be?

The final issue is the effects of global warming and increased CO_2 levels. We have already seen that the earth has been warmer than it is now, possibly even during the lifetime of the human species. People experience CO_2 levels twenty or more times the atmospheric norm in crowded rooms every day. CO_2, even well in excess of the current level, is not poisonous to humans. To label it a pollutant represents a political rather than scientific decision by the EPA.[26]

Mankind can also survive higher temperatures. Ilulissat is perfectly happy to see the glaciers melting and the sea warming. Siberia and Alaska probably feel much the same. On the other hand, change is always disruptive, so low-lying settlements near the sea may be at risk. But of course, the people living in the Netherlands continue to adapt to sea levels that have risen for centuries. And London in 1982 erected the Thames Barrier to protect the city from increasingly high tides; Venice, in the fullness of time, will likely do the same. There are ways to cope with rising sea level. The effects of temperature changes will surely be felt slowly enough to allow us to adapt. But the rapidity of climate change and its unmanageability without radical changes in society are simply taken for granted by alarmists.

We also know that increased atmospheric CO_2 usually aids plant growth.[27] Even NASA, which accepts the AGW hypothesis, acknowledges this, albeit rather grudgingly, headlining that "CO_2 Is Making

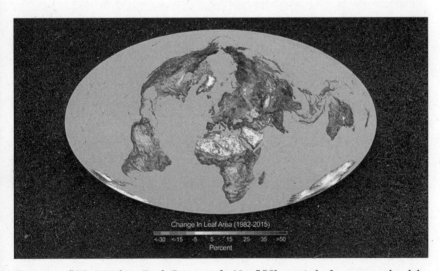

FIGURE 10. "CO_2 Is Making Earth Greener—for Now." "Change in leaf area across the globe from 1982–2015." Source: Samson Reiny, NASA's Earth Science News Team, April 26, 2016, https://climate.nasa.gov/news/2436/co2-is-making-earth-greenerfor-now/.

Earth Greener—for Now."[28] Figure 10 shows the satellite-measured change in global leaf area between 1982 and 2015. According to NASA: "A quarter to half of Earth's vegetated lands has shown significant greening over the last 35 years largely due to rising levels of atmospheric carbon dioxide, according to a new study published in the journal *Nature Climate Change*."[29]

Will this benefit persist, or will it be limited by other factors, as some have suggested? Are there other benefits to warming? After all, more people seem to die of harsh winters than hot summers.[30] Perhaps more time should be spent studying the potentially beneficial effects of CO_2 and less time trumpeting an uncertain apocalypse. There are at least as many reasons for optimism as the reverse. It is hard to justify the end-of-days alarmism that now consumes so many. Perhaps the collective passion is just groupthink, a contemporary example of the madness of crowds and a reminder that a group is not always wiser than a thoughtful and disinterested individual.[31] Alarmism about climate change does not seem to follow from science.

PART 4

Social Science

CHAPTER 10

Killing the Messenger

"A scientific man ought to have no wishes,
no affections, a mere heart of stone."

—*Charles Darwin*

Social science is almost an oxymoron. There are two reasons. The first is that much of it is what physicist Alvin Weinberg many years ago called trans-science. Weinberg's distinction between science and trans-science seems to have been forgotten.[1] Trans-science refers to questions that are scientific but for practical or ethical reasons cannot in fact be decided scientifically. Human behavior, both individually and in society, is exceedingly complex, and a meaningful experiment is very often impossible. Scientific questions can be posed, but often they cannot be answered. For example, for ethical and practical reasons the effect of low-level pollutants on kids' mental ability simply cannot be decided by experiment. You can't expose children to something that may harm them; you don't really know what to use as a measure of mental ability; and you don't really know the appropriate time frame: Do you look at their behavior in a year? Ten years? Until middle age? In the absence of fundamental knowledge of causal mechanisms—in this case the physiology and biochemistry underlying the effects of the pollutant in question—we are left with correlations that cannot prove causation, or theoretical extrapolations that are often unwarranted. The spread of COVID-19 was ill-predicted by models that could not hope to capture the millions of interactions that promote the spread of a new virus with many unknown properties. But people wanted answers, and so answers, many of them wildly wrong, were provided—and the reputation of "science" took another hit.

The trans-science problem won't go away, but if we recognize the limitations on what can actually be achieved, more modest questions can be asked, and something can be learned.

The second reason that social science often fails is more intractable: it is the emotional impact of many scientific questions, questions that should be "value neutral," as I explained earlier. Emotion rules out research on many topics and many answers to questions on permitted topics. If those questions cannot be asked, much less answered, future understanding will be warped in unknown ways. We will think we know things that we do not, and we will fail to recognize causes we have intentionally ignored.

Here is a small-scale example from my own institution of how even a simple, matter-of-fact, empirical result may inflame a susceptible and ill-educated audience.

Racial Differences?

A 2011 statistical study coauthored by two Duke University economists and a sociologist, and still unpublished at the time, raised a ruckus. Here is the abstract:

> At [Duke University], the gap between white and black grade point averages falls by half between the students' freshmen and senior year ... [which suggests] *that affirmative action policies are playing a key role to reduce racial*

differences. However, this convergence masks two effects. First, the variance of grades given falls across time. Hence, shrinkage in the level of the gap may not imply shrinkage in the class rank gap. *Second, grading standards differ across courses* in different majors. We show that controlling for these two features virtually eliminates any convergence of black/white grades. *In fact, black/white gpa convergence is symptomatic of dramatic shifts by blacks from initial interest in the natural sciences, engineering, and economics to majors in the humanities and social sciences.* We show that natural science, engineering, and economics courses are more difficult, associated with higher study times, and have harsher grading standards; all of which translate into *students with weaker academic backgrounds being less likely* to choose these majors. Indeed, we show that accounting for academic background can fully account for average differences in switching behavior between blacks and whites [emphasis added].[2]

One reason the study became a *cause célèbre* at Duke was because it had recently been involved in a case against affirmative action in the U.S. Supreme Court.[3] The study provides supporting evidence: affirmative action almost guarantees the admission of a racially defined cohort who would not have made the grade under the standard policy. For this group, race weighs more and academic qualifications less than other admits. It should not surprise that in some comparisons they may on average perform worse, a fact which might well be used as part of an argument against affirmative action.

There are other results in the study; many analyses were done. But the switching-major result seems to be the one that raised most hackles. Whites do quite a bit better than blacks during freshman year, but this difference almost disappears for seniors. The authors argue that switching majors, rather than improvement in black students' performances, accounts for the reduction in grade-point-average difference between black and white students that occurs between freshman and senior years.

Why switch major? Switching seemed to reflect not race but "academic preparation," since the study found a similar pattern for legacy students,[4] admitted, like African Americans, according to somewhat different criteria than the average white or Asian applicant.

In sum:

1. Entering black students prefer STEM courses more than do whites.
2. But they tend to shift their preference to social science and humanities in later years; white students not so much.
3. The difference between white and black average GPA declines from 0.5 (3.38–2.88) to 0.33 (3.64–3.31) between students' freshman and senior years.
4. Inference: this isn't so much because the black students improve their performance but because they switch from harder STEM courses to easier social science and humanities courses.
5. Legacy students follow the same pattern.

Point 4 might be questioned. Quite possibly some students switch majors simply out of interest: having sampled both STEM and social science courses, say, they find they like the social science courses better. The authors are in fact noncommittal on this point, merely saying that the tougher grading standards in science courses make students with "weaker academic backgrounds . . . less likely" to take them.[5] Otherwise, the results seem to be straightforward. Their generality remains to be established. Is this pattern true of other elite schools? Will it still be true of Duke in a few years? Never mind, for now we have one data point; more research may be needed.

It's hard for many people to see what the big deal is in this study. Science is tougher than nonscience? "The fact that sciences are harder than humanities is the worst kept secret in higher education," wrote a student to the *Duke Chronicle*. Students tend to switch out of subjects in which they are doing badly. No kidding!

Yet this totally factual, common-sense academic study caused some consternation at Duke, including a protest by the Black Student Alliance, which featured placards like "Does GPA Have a Color?" "'This study does not embody Duke's values as an institution,' said [a sophomore student] who attended the demonstration. 'We do not stand for that type of racist inquiry and that misuse of academia to mischaracterize the accomplishments of the African-American students at our institution.'"[6] The student did not explain just how these accomplishments were "mischaracterized" or why the study is "racist"—or indeed, to what "Duke values" he is alluding. The Black Student Alliance also rejected the obvious fact that humanities and social sciences are easier for most students than STEM courses, inadvertently boosting the very stereotype to which they object: that blacks are not as smart as whites.

Never mind the details. The point is that facts—about how different groups of students choose and switch courses, about average GPA differences—led to angry criticism, not of the facts or even of how they were collected, but criticism of their very existence. Yet it is obvious that affirmative action, which favors one racial group over others in a world where the favored group scores, on average, below the rest, *must* lead to differences in performance between those groups. These differences are almost certain to lead to the kinds of things reported in this study. The students' solution was to urge that studies like this be in effect banned at Duke.

Duke's president responded in his annual address to the faculty. After deploring Duke's past "policy of racial exclusion" while recognizing that "the founding family of Duke University was extraordinarily progressive for its time in the matter of race," he sympathized with the students, referring to "faculty research that appeared to disparage the choice of majors by African-American undergraduates," as if reporting students' choices was the same as disparaging them. "I can see why students took offense at what was reported of a professor's work," he concluded.[7]

What no one pointed out was that these students (or any university students) were wrong to get fired up over a fact. Facts are what education is about. Criticize them, the method used to get them, or the inferences made from them; but facts are, or should be, just facts. The distraught students were treated like infants. They were humored, pandered to, conciliated—not educated. And the cry for censoring this kind of research was tolerated rather than refuted. This is now the prevailing pattern in academe.[8]

Censoring Science

Censorship like this has several sources:

- The most important is that the data point to an unacceptable conclusion.
- The research deals with a taboo subject, like race differences.
- But this factor is applied *inconsistently*.

- It is acceptable to study race, say, so long as the results identify negative causes over which the individuals have no control, particularly racism, systemic or otherwise.
- The research is not acceptable if the results implicate causal variables over which individuals do have some control, like family structure, academic performance, and personal responsibility.

How to fix the problem? First, by affirming that the *only* legitimate response to a scientific report is a scientific one: criticize the method, or the inferences the authors draw from their results. What is never legitimate, if science is to survive as a way of understanding nature, is to condemn and seek to exclude both the facts themselves and any interest in the topic.

Researchers should not have to worry about being called on the carpet any time their work might cause anyone to take offense. The solid ground of academic freedom starts turning into quicksand when university presidents rebuke faculty members for drawing "insulting" conclusions or "disparaging" some group simply by presenting a fact that may upset some people. Killing, or at least silencing, the messenger will mean the death of science.

The Devolution of Social Science: How the Fragmentation of Sociology Has Led to Absurdity

Some Science History

At the dawn of science, practitioners were few and all had some acquaintance with every branch. In the original Royal Society of London (founded in 1660), for example, papers were presented before the whole group, and everyone felt free to comment on and evaluate what they heard. There were no well-defined subdisciplines; science—or *natural philosophy*, as it was then called—was not a *profession*, like law or medicine.

Most scientists did serious work in many areas: Isaac Newton (Royal Society president, 1703–1727) did mathematics, optics, astronomy—and astrology and alchemy—as well as what is now termed "physics." Christopher Wren, architect by profession, was also an anatomist, geometer, astronomer, and physicist. Edmund Halley, of "Halley's Comet" fame, made contributions in mathematics and optics as well as astronomy. Halley could comment competently on an analysis of the credibility of human testimony or a perceptual phenomenon like the moon illusion.[1] There was ample opportunity for any interested party to criticize or comment on any piece of work. None could be dismissed as inexpert or a nonspecialist.

The British Association for the Advancement of Science (now the British Science Association, or BSA), founded in 1831, had by 1873 divided into several sections: chemistry and geology, zoology, and physiology and geography. In the 1880s, engineering, and two somewhat-social sciences, economic science and statistics, and anthropology, were added. But originally the entirety of physics and mathematics was in Section A, which covered a wide range of subjects: "I have heard a discussion on spaces of five dimensions and we know that one of our committees . . . reports to us annually on the rainfall of the British Isles," remarked one commentator.

The diversity of topics led to suggestions that Section A be subdivided. But even in 1873 there was resistance "instancing the danger of excessive specialization, and claiming that the bond of union among the physical sciences is the mathematical spirit and mathematical method."[2] Nevertheless, specialization prevailed. By 2018 the British Science Association had developed seventeen sections.[3]

Social science, restrained neither by a common theory nor a common method, has been less resistant to subdivision. When the American Psychological Association was founded in 1892 it had just one division, but after World War Two it merged with various other psychological organizations and created nineteen divisions. By 2007, this number had expanded to fifty-four; in the meantime, a competing entity, the Association for Psychological Science (APS), had split off from the American Psychological Association (APA), and APA Division 25 had itself given birth to an independent organization of its own called the Association for Behavior Analysis International (ABAI).

Parallel to the American Psychological Association and with a smaller membership (circa 13,000 versus 70,000 plus)—but with almost as many sections (fifty-three)—is the American Sociological Association (ASA).

The reasons for this fissiparousness are more social than scientific. Each area tends to define its subject matter a bit differently, even though to outsiders it looks as if all are in fact studying much the same thing. Methods thought legitimate in one division are deemed inadequate in another, and so on. In psychology, for example, behaviorists and cognitivists could not agree on the proper object of study: is it *behavior* (the behaviorists) or *mind* (the cognitivists)? The result: APA Division 25 and the Association for Behavior Analysis International for the behaviorists, and many other divisions of the American Psychological Association and most of the Association for Psychological Science for the cognitivists. Having subdivided in this way, and fortified their different camps, the two groups could pursue their different approaches without conflict and *with little chance those differences will ever be resolved.* The same is true for the multitude of other groups that now exist.

A fatal consequence of this multiplication of social-science subdisciplines has been a weakening of criticism. Technical languages—mostly jargon—have evolved and, like natural languages, isolate speakers from nonspeakers, immunizing research in one area from truly independent criticism from workers in other, related areas. The result has been the emergence of a wide range of new specialties, some truly creative, but many bearing little resemblance to "hard" science. The best-known example of this erosion of scientific standards is the ongoing replication crisis in social and biomedical science. The real problems go well beyond replication, however.

As practical applications of science continued to grow during the nineteenth century, resources began to flow into the scientific enterprise. Biological and then social topics began to be studied from a self-consciously scientific point of view. As the number and diversity of practitioners and professional subdivisions increased, so did the numbers of scientific publications. Each scientific society had at least one journal and usually several. Now, the number of scientific journals has increased to the point that no one seems to be sure of their precise number, except that it is very large indeed. The best estimate as of 2018 is between twenty-five and forty thousand. The number of published papers each year is of course larger still.[4]

In social science, especially, each subdivision developed its own cluster of journals. A finding in one area that might have little or no credibility in others, can nevertheless find a safe berth in its own publication harbor. Results can be supported by citations drawn from a sympathetic group of like-minded researchers, often publishing in the same journal, in a circular support group.

Sociology as a Science

This unchecked subdivision has allowed sociology to take seriously the mysterious and paradoxical notion of color-blind racism, discussed in the next chapter. If the following argument is hard to follow, the fault is only partly mine. Modern sociology bounces merrily along from science to anecdote, and from storytelling, to propaganda, to activism—often within the same paragraph. I have tried to separate these themes and to differentiate what is scientific from what is merely intended to persuade or induce social change, but it isn't always easy.

Sociology originated in a self-consciously scientific way: French scholar Émile Durkheim (1858–1917), one of the founders, explained his purpose:

> Our main objective is to extend the scope of scientific rationalism to cover human behaviour by demonstrating

that, in the light of the past, it is capable of being reduced to *relationships of cause and effect*, which, by an operation no less rational, can then be transformed into rules of *action for the future* [emphasis added].[5]

Cause-and-effect and rational analysis, the basics of science, were to be at the heart of this new social science. But also rules of *action*—an apparent afterthought by Durkheim that has turned out to be far from benign.

Durkheim was well aware that sociology touched on many other fields, from psychology to history, economics to anthropology. Hence, he sought to define what he called *social fact*, a kind of fact that lies outside the facts of any other social science. He first acknowledged the difficulty of the problem:

[The term "social fact"] is commonly used to designate almost all the phenomena that occur within society, however little social interest of some generality they present. Yet under this heading there is, so to speak, no human occurrence that cannot be called social. Every individual drinks, sleeps, eats, or employs his reason, and society has every interest in seeing that these functions are regularly exercised. If therefore these facts were social ones, sociology would possess no subject matter peculiarly its own, and its domain would be confused with that of biology and psychology.[6]

Durkheim goes on to claim that in fact "there is in every society a clearly determined group of phenomena separable, because of their distinct characteristics, from those that form the subject matter of other sciences of nature" and gives an example:

> When I perform my duties as a brother, a husband or a citizen and carry out the commitments I have entered into, I fulfil obligations which are defined in law and custom and which are external to myself and my actions. . . . Considering in turn each member of society, the foregoing remarks can be repeated for each single one of them. Thus there are ways of acting, thinking and feeling which possess the remarkable property of existing outside the consciousness of the individual.[7]

These are the facts of sociology—outside the individual consciousness but affecting people nevertheless—moral rules, the language spoken, the legal currency, the mode of dress, "a category of facts which present very special characteristics: they consist of manners of acting, thinking and feeling external to the individual, which are invested with a coercive power by virtue of which they exercise control over him."[8] Durkheim laid out a subtle path for his new science, related to, but separate from, all other human sciences. Modern sociology bears little relation to Durkheim's plan; some examples follow.

Causes and Social Facts

Social facts in Durkheim's sense are very hard to find. Identifying causes is even more difficult. Nevertheless, most contemporary sociologists give lip-service to Durkheim's scientific ideal. For example, Eduardo Bonilla-Silva, a leader in the color-blind racism (CBR) movement, in the fourth edition of his landmark book *Racism without Racists*, cites approvingly a "biting critique of statistical racial reasoning" by Tukufu Zuberi, now a distinguished professor of sociology at the University of Pennsylvania.[9] The reference leads the reader to expect a sophisticated analysis of social causation. And indeed, Zuberi's article affirms, in a roundabout way, that statistical correlation is not the same as causation. But then he quotes approvingly another author who says "the schooling a student receives can be a cause, in our sense, of the student's performance on a test, whereas the student's race or gender cannot"—a peremptory dismissal of race or gender as part of any explanation of differences in test performance.[10]

This claim ignores the difference between efficient and material causes, a distinction that has been around at least since Aristotle.[11] When you turn on your television, a picture appears. The switching-on was the *efficient* cause. But the machinery of the television set and the whole transmission network is hardly irrelevant: it constitutes the *material* cause. The term *state* is also used to describe this kind of cause,[12] as in "the picture will only appear if the TV is in the right state, that is, plugged in, not broken or damaged."

What constitutes a cause, and what kind of cause is constituted, depends on the subject and level of analysis. Every material cause, like the state of the television, is itself an effect of prior efficient cause(s), such as the manufacturing process. The capacities of an individual, his constitution and his memory, are the *material* (state) cause of performance on a psychological test. The test questions are the efficient causes of the subject's responses. Education may be an efficient cause of test performance, as Zuberi says. But it acts through the constitution of the pupil, a material cause. And the pupil's constitution is in turn the effect of genetic and developmental processes.

Why the Error?

Zuberi's failure to recognize a material cause is puzzling because it is so elementary. Why did he make it, and why did the journal editors overlook it? We can perhaps understand the reason for the error if we look at its effect, which is to explain a student's academic performance purely by exogenous factors like (usually inadequate) education and racial discrimination. Endogenous factors—whether he works hard or not, is clever or not, possible effect of the student's home environment, and so forth—are thereby elided. Society is to blame for any problems, not the individuals who display them or even the homes from which they came. In short, the epistemic error has a political effect.

But why did the journal editors not demand that such an elementary error be corrected? The rest of Zuberi's article provides clues. Much attention is paid to the work of scientific pioneer Francis Galton (1822–1911). Galton is famous for many things, but one thing for which he is now infamous is *eugenics*,[13] the idea that humanity should use our knowledge of heredity to encourage the birth of improved versions of ourselves: not so much "better living through chemistry" as "breed better babies." In the first part of the twentieth century the idea was accepted as no more than commonsense by many Western intelligentsia—until the Nazis gave it a very bad name indeed (not that Hitler even used the word in *Mein Kampf*).

Galton believed that individuals, classes, and races differ in their endogenous abilities. As Zuberi says,

> Galton used statistical analysis to make general statements regarding the superiority of different classes within England and of the European-origin race, statements that were consistent with his eugenic agenda.[14]

Galton's name is mentioned eleven times and "eugenics" five times in Zuberi's article, alerting editors to the danger of giving these ideas any credence. They got the message and Zuberi's elision is given a pass. A small failure of logic, but a big shift in attention.

The point of this rather labored example is to show how in sociology politics can infect supposedly scientific judgement. Politics says that black-white differences in average IQ are solely caused by the social and educational deprivations of blacks, and not by *endogenous* group differences between blacks and whites. Any research evidence that implies otherwise is simply taboo.[15] So, what should have been a straightforward causal analysis is adjusted accordingly.

The next chapter shows how the devolution of sociology has led to unreason, in the form of the subdiscipline of race and ethnic studies and the concept of color-blind racism (CBR)—the idea that treating people according to the content of their character, not the color of their skin, is itself racist.[16] Martin Luther King's famous definition of nondiscrimination is rejected by, for example, the 2018 president of the American Sociological Association.

CHAPTER 12

Contemporary Sociology:
Race and Ethnic Studies

T he previous chapter showed how a branch of sociology has been able to narrow the definition of causation. In what might be called *race-and-ethnic-studies* (RES) sociology, efficient (exogenous) causes rule; material (endogenous) causes are excluded. What about Durkheim's elusive social facts? Perhaps the most famous candidate is a contribution of his equally famous contemporary Max Weber's idea of the *Protestant (work) Ethic* as a major cause of the success of capitalism.[1] Weber's argument is closer to economic history than the kind of subtle social analysis envisaged by Durkheim. But Weber also analyzed bureaucracy in terms of its structure (division of labor), regulatory hierarchy, and meritocratic hiring practices. These factors seem closer to Durkheim's idea, being both measurable as well as extra-individual.

Things are different now; "fact" itself has become a suspect idea. I first got an inkling of this more than three decades ago. Sorting through some old academic papers, I found this quote from an unnamed British sociologist speaking at a talk in 1986: "Theories in science are not constrained in any way by empirical facts." I noted that most of those listening agreed with him.

The quote is absurd, but in the years that followed I noted how widespread this assault on the scientific method has become. A whole field devoted to diminishing if not discrediting science as a source of objective knowledge has sprung up under the banner of "Science Studies," which is now a recognized academic discipline with its own association and cluster of peer-reviewed journals. Science studies has several branches, from reasonable (the study of reactions to natural disaster, the role of probability in science, for example) to others not so rational (scientific knowledge as "socially constructed," the illusion of objectivity, et cetera). One of the less rational is the journal *Social Text*, which a few years ago published a brilliantly nonsensical hoax piece, "Transgressing the Boundaries: Towards a Transformative Hermeneutics of Quantum Gravity," by physicist Alan Sokal.[2] Sokal succeeded by using favored jargon (like "transgressive," "problematize," and "hegemony"), promoting the correct political views (like that which views science as gendered domination), and putting "objective" in quotes.

The anonymous sociologist's claim that empirical facts are irrelevant does apply to much of contemporary social science. It raises an obvious question: If theories in the social sciences are not constrained by empirical facts, what *are* they constrained by? The answer seems to be that theories in areas like RES sociology are mainly constrained by the political ideology prevailing in that branch of the field.

Race and Ethnic Studies

In RES it is simply assumed that the findings and reasoning of sociologists are determined by their ethnicity and position in society. Editors Zuberi and Bonilla-Silva in a book called *White Logic, White Methods: Racism and Methodology* write:

Most White sociologists, reflecting their dominant position in the discipline, have complained that sociologists of
color are "biased" and thus do not take seriously their work
or their criticisms. Conversely, many sociologists of color,
reflecting their subordinate position in sociology, have
doubted the research findings by white sociologists to
explain the standing of people of color in America.[3]

This passage is just one of many that either directly or indirectly
denies the possibility of objectivity, hence excluding RES scholarship
from science. It perhaps also explains why the authors don't bother
to present any empirical evidence in defense of their ad hominem
identity epistemology. We're just supposed to accept its truth without
question, although, having said that, the concept of "truth" appears
to be equally suspect. Indeed, in another place, Bonilla-Silva scorns
the very idea, speaking of the "devil of 'objectivity'" (which obviously
does pose problems for the RES movement):

We made a pact with the devil of "objectivity." Our pact,
made in earnest in the 1920s, has become normative . . .
We still believe the impossible: that we are above the social
fray. We believe that we should not be "political" and
assume our job is to adjudicate, based on "data" and
immaculate methods, who is right or wrong on social matters. *But all this is nonsense. . . .* I for one believe we can be
passionate and committed to social change, yet be rigorous
in our work [emphasis added].[4]

It is not necessary to believe we are already perfectly objective—
"above the social fray" in Bonilla-Silva's words—in order to strive for
objectivity, any more than it is necessary to be perfectly healthy to

strive for health. But what if you are "committed to social change" and the truth points in another direction? Suppose the data point to endogenous rather than exogenous causes for a racial disparity, will you accept the truth or ignore it? Without an honest commitment to objectivity, RES sociology ceases to be science.

Has RES sociology become, then, just political activism? To some extent, yes. According to Zuberi and Bonilla-Silva, their aim is "to attain epistemic liberation from White logic," and they add that "we see this edited volume as part of the *long march of resistance* to White domination in society and in academe [emphasis added]."[5] The Gramscian allusion is probably not accidental.[6]

By the end of the book Zuberi and Bonilla-Silva have backtracked slightly, making a shallow bow to objectivity while echoing the earlier ad hominem: "Rather than leading to a science of objectivity, White logic has fostered an ethnocentric orientation. . . . however, scholars of color are potentially much closer to being objective . . ."[7] This will leave many readers puzzled: Is the work "biased" when the sociologist is white—or, rather, "White," to add the mandatory scare quotes—but objective when she is a person of color? The authors attempt to clarify by quoting Charles W. Mills:

> Hegemonic groups characteristically have experiences that foster illusory perceptions about society's functioning, whereas subordinate groups characteristically have experiences that (at least potentially) give rise to more adequate conceptualizations.[8]

So the worm's-eye view is more "objective" than the bird's-eye view—or, to use the jargon, apparently "subordinate" groups (for example, people of color) see things more clearly than "hegemonic groups." Since Jamaican-American Mills presumably considers himself a member of a subordinate group (even though he is a distinguished professor of philosophy in the City University of New York Graduate Center) his claim of subordinate superiority invites the Mandy Rice-Davies response: "Well, he would say that, wouldn't he?"[9]

Social Facts—RES Style

Here are some ideas that are generally accepted by the race-and-ethnic-studies movement as social facts (although the movement's allegiance to truth is shaky, as we have seen):

White Logic

White logic is the idea that white people think differently than people of color. It is embedded in "the structure that generates racism," in the words of Anna-Esther Younes, reviewer of Bonilla-Silva and Zuberi's *White Logic*. She concludes that "Ultimately, what connects all authors is their view of academia as 'a form of [White] cultural and political hegemony.'"[10] (*Hegemony* is a popular term in the RES literature. It seems to mean "traditional ideas which I deplore.")

Zuberi and Bonilla-Silva acknowledge that many readers will find the topic distasteful:

> "Why did you folks write a book on White logic and methods?" They will likely be incensed . . . The methodologically inclined will say, "Methods are objective research tools beyond race, gender and class." They will argue that "social science methodology, like genetics, can be applied impartially regardless of the racial background of the individual conducting the investigation."[11]

Well, yes, they will; so what is the authors' response? "Before we address these burning questions, we need to explain our motivations for editing [this] book."[12] And from Bonilla-Silva's *Racism without Racists*: "At every step of the way, I have encountered people who have tried to block my path one way or another."[13] He goes on to list literally dozens of people who have in fact helped him and then later writes: "How can I, after been [*sic*] elected president of the American Sociological Association (ASA), be talking about the salience of race in our business? Isn't my election proof positive that race is 'declining in significance' . . .?"[14] The obvious answer to this question is yes. Why should we answer differently?

In fact, we *never get an answer* to these questions. Just as the authors seem about to define a term or justify a claim, the football is pulled away, to be followed by repeated assertions about links to colonialism or white feelings of superiority. The closest we get to a definition of white logic is Zuberi's answer to the hypothetical: "Are you suggesting social scientists practice racism when they use statistics?"[15] His conclusion, although he never says so directly, is yes. His reason is Francis Galton, Darwin's half-cousin and a founder of the statistical method. Galton "was obsessed with explaining racial hierarchy in social status and achievement."[16] Well, yes, Galton was interested in what made for success in life. In his 1869 book *Hereditary Genius: An Inquiry into Its Laws and Consequences*, he says that the "negro race" is an "inferior race," but he also says that "the average ability of the Athenian race is, on the lowest possible estimate, very nearly two grades higher than our own—that is, about as much as our race is above that of the African negro."[17]

He goes on to say:

> There is nothing either in the history of domestic animals
> or in that of evolution to make us doubt that a race of sane
> men may be formed who shall be as much superior

mentally and morally to the modern European, as the modern European is to the lowest of the Negro races.[18]

If Galton was racist, he was evenhanded, and by no means biased in favor of modern Europeans. And in every case, he made a coherent argument based on the best evidence available to him. Galton knew nothing of the still-controversial genetics of white and black "races," for example,[19] and those few African blacks he may have encountered probably came from pre-literate, rural cultures and spoke little English. Intelligence tests had not been invented, but there we may be certain that these Africans would not have done well on them.[20] Galton's take may well have been perfectly reasonable, based on the information available to him. It is just silly to claim that "current statistical methodologies . . . continue to reflect the racist ideologies" of the eugenics movement.[21]

As for eugenics—for his interest in which, Galton is banished from most of social science—the controversy is about political action, not scientific understanding.[22] As Galton says, there is no doubt that the human race, like the race of any animal, could be changed, even improved, by selective breeding. No biologist would disagree with this. But whether it should be attempted, and if so, how, by whom, to what end, and under what authority—all this involves judgments that are ethical and political, not scientific. Science is about *is* not *ought*, as I have pointed out repeatedly. Eugenics is a political proposition not a scientific one.

Whiteness

Whiteness, "the practices of the 'new racism'—the post-civil rights set of arrangements that preserves white supremacy," in the words of Bonilla-Silva[23]—is apparently hegemonic: "I contend that 'color-blind' ideology plays an important role in the maintenance of white hegemony," writes Ashley "Woody" Doane, a leading "whiteness studies" advocate who heads the sociology department at the University of Hartford.[24] "Whiteness" is employed as a method of maintaining control over other groups by the "dominant culture." Hence, "challenging white hegemony" is a major motif for "whiteness studies." According to Bonilla-Silva, only "'race traitors'—whites who do not dance to the tune of color blindness"—can escape from whiteness.[25] Color blindness is part of the whiteness strategy and is therefore racist.

Here is a description of "Whiteness Studies" from George Yancy, a leader in the field, whose interviews appear frequently on the pages of the *New York Times*. On a now-vanished website he describes his own work as follows (referring to himself in the third person): "Yancy explores the theme of racial embodiment, particularly in terms of how white bodies live their whiteness unreflectively vis-a-vis the interpellation and deformation of the black body and other bodies of color."[26]

Many readers will find this account hard to follow. Yancy's dependence on jargon is characteristic of many of the subdisciplines of social science and humanities. The terms are almost never clearly defined, but they serve at least two purposes: they convey expertise—and impede scrutiny.

This kind of academical obscurantism is not new. Thomas Hobbes in *Leviathan* (1561) complains about "names that signifie nothing; but are taken up and learned by rote."[27] Theological writing is perhaps clearer now, but the arcane rhetorical techniques that provoked Hobbes's ire have not been lost.

Subjective experience—private feeling, *Erlebnis*, known to philosophers as *qualia*[28]—is obviously very important to whiteness studies and, indeed, to sociology generally (the theme of the 2018 national conference of the ASA was *Feeling Race*). The problem is this: qualia cannot be measured by a third party and so lies outside of science.[29]

Subjective experience can be shared through drama or literature—or simply storytelling, which is a major part of RES literature. But science is about *public*—third-party-accessible—data. Physicist John Ziman expanded on this idea in a brilliant book and used it as a definition of all science, which he called "public knowledge."[30] The varieties of cognitive and behavioristic psychology agree.[31] Qualia are not part of science and neither is RES sociology.

"Whiteness," like color-blind racism, is also unconscious. A former student of Bonilla-Silva's raises the obvious question: "How does one test for the unconscious?"[32] But, like Bacon's Jesting Pilate, Bonilla-Silva stays not for an answer. Others have tried. There is something called the "implicit-bias" test which pretends to measure unconscious processes. But it is neither reliable nor valid; it has little or no scientific basis.[33] Despite their fallibility, such tests have been widely administered for some twenty years, and there is still a Harvard University website that hosts one. It will be job for future sociologists to figure out the reasons for their success.

It is also reasonable to ask why a test for the *unconscious* is relevant to anything other than neuroscience (perhaps different qualia are associated with different brain states?). In a Western democracy without enforced mind-control, only actions matter. If the implicit-bias test does not predict explicit bias (and it does not) it serves only as employment for diversocrats and a weapon for activists.

To many modern sociologists, color blindness is a racist creed that works, somehow, through whiteness, a scheme of thought invisible to most whites, but revealed by RES sociology. Whiteness is part of systemic racism: "Exposing the Whiteness of Color Blindness" is a chapter subhead in Bonilla-Silva's book.[34] Whiteness is as real an identity as blackness. None of these, neither whiteness, nor blackness, nor systemic racism is measurable in an objective way. Given the many failures of logic in this literature, one reading is that "whiteness" is just another name for "reason."

Privilege

Above all, in the RES universe whiteness is a bearer of *privilege*. Privilege is the flip side of discrimination: to discriminate against *A* is to privilege not-*A*. Discrimination against blacks implies unfair advantage—privilege—for whites. The term is therefore redundant. It originated with sociologist W. E. B. Du Bois but has recently been revived. As evidence for actual racial discrimination has become harder to find, *privilege* remains a way to explain racial disparities—and discomfort whites. Books and articles in this area are sprinkled with tendentious, not to say racist, phrases like "the manifold wages of whiteness," "white privilege," and "historically white colleges," all to emphasize persistent, unjust advantages possessed by whites as opposed to blacks. Again, the injustice of privilege is just assumed, not demonstrated empirically. The few demonstrable examples of "*black* privilege," such as affirmative action and near-universal diversity policies—real systemic racism—are ignored.

A 2014 interview with Peggy McIntosh, a women's studies scholar at Wellesley, gives the history of the term.[35] McIntosh in 1988 circulated a list of forty-six "privileges" accorded to white people. The piece, entitled "White Privilege and Male Privilege: A Personal Account of Coming to See Correspondences through Work in Women's Studies," turned out to be a big hit. It included privileges such as "I can if I wish arrange to be in the company of people of my race most of the time," "I can turn on the television or open to the front page of the paper and see people of my race widely represented," "I can be pretty sure that if I ask to talk to the 'person in charge,' I will be facing a person of my race," "I can take a job with an affirmative action employer without having my co-workers on the job suspect that I got it because of my race," and many more.[36]

Some of the listed privileges reflect the fact that blacks are a minority in America, some reflect housing segregation, voluntary and otherwise, and some, like affirmative action, reflect government policy (vigorous affirmative action means you may well have got the job because of your race). The privileges make tacit, but slightly contradictory, assumptions: that blacks and whites are equivalent in terms of both interests and abilities, and yet wish to congregate separately. Basically, what McIntosh is saying is that blacks more often think about their race than whites do about theirs, which is probably true for many minorities. McIntosh's piece makes entertaining reading, but as science, or proof that privilege is something other than the obverse of racial discrimination, it is a nonstarter. Nevertheless, the term "privilege" has become popular and McIntosh has found a receptive audience of sympathetic, white elites.[37]

White Fragility

Robin DiAngelo is an Affiliate Associate Professor of Education at the University of Washington, Seattle. Her "area of research is in Whiteness Studies and Critical Discourse Analysis, tracing how whiteness is reproduced in everyday narratives."[38] She also has a bestseller: *White Fragility: Why It's So Hard for White People to Talk about Racism*, which sociology professor Michael Eric Dyson calls "vital, necessary, and beautiful."[39] DiAngelo has been giving dozens of well-paid talks about white fragility to nationwide audiences.[40] Her book apparently "allows us to understand racism as a practice not restricted to 'bad people.'"[41] A gushing *New Yorker* reviewer finds it to be a "methodical, *irrefutable* exposure of racism in thought and action [emphasis added]."[42]

It is tough to classify a research area like this, which clearly aims to persuade rather than enlarge scientific understanding. But because DiAngelo's work is influential and she cites sociological sources, I will treat it as social science, even though her writing is little more than a set of ill-defined and unsupported allegations that fit more easily into some branch of advertising or a contribution to Lifehack.org than any kind of science.

DiAngelo begins a much-lauded paper on white fragility with a series of allegations:

> White people in North America live in a social environment that protects and insulates them from race-based stress. This insulated environment of racial protection builds white expectations for racial comfort while at the same time lowering the ability to tolerate racial stress, leading to what I refer to as White Fragility. White Fragility is a state in which even a minimum amount of racial stress becomes intolerable, triggering a range of defensive moves. These moves include the outward display of emotions such as anger, fear, and guilt, and behaviors such as argumentation, silence, and leaving the stress-inducing situation. These behaviors, in turn, function to reinstate white racial equilibrium. This paper explicates the dynamics of White Fragility.[43]

This is followed by an italicized (apparently autobiographical) paragraph: "I am a white woman. I am standing beside a black woman. . . . I have just presented a definition of racism that includes the acknowledgment that whites hold social and institutional power over people of color. . . ." Apparently a white man gets upset. DiAngelo is puzzled: "We have, after all, only articulated a definition of racism."[44] A definition that contains an unproved accusation against white people. Surely a reaction from a white person is hardly unexpected.[45] Are his objections justified? We don't know and DiAngelo has no interest in the question. Quoting Michelle Fine, she continues with more allegations:

"Whiteness repels gossip and voyeurism and instead demands dignity." . . . This insulated environment of racial privilege builds white expectations for racial comfort while at the same time lowering the ability to tolerate racial stress. . . . White people enjoy a deeply internalized, largely unconscious sense of racial belonging in U.S. society.[46]

DiAngelo confidently attributes unconscious motivations to white people. But psychologists continue to struggle with the role of the unconscious.[47] She seems to be completely unaware of the uncertainty surrounding the concept and the enormous difficulties psychologists have in measuring unconscious effects. (See, for example, the difficulties encountered by the so-called implicit-bias test mentioned earlier.)

Whites are said to be "stressed." What does this mean? How is stress measured or validated? How is "race-based stress" different from other kinds of stress? "This insulated environment of racial privilege builds white expectations for racial comfort while at the same time lowering the ability to tolerate racial stress." Really? How is "racial comfort" measured? How does the process lower the ability to tolerate racial stress? None of these claims is followed by proof, other than citations of other authors who just make similar, unsubstantiated claims.

There's more: "Whites are theorized as actively shaped, affected, defined, and elevated through their racialization and the individual and collective consciousness' formed within it."[48] "Theorized"? Does that mean "hypothesized"? In which case, where is the test? "Whiteness Studies begin with the premise that racism and white privilege exist in both traditional and modern forms, and rather than work to prove its existence, work to reveal it."[49] How are the "traditional and modern forms" of "white privilege" and "racism" defined? Where is proof of the premise? How can you "reveal" something whose existence is yet to be proved?

What on earth is a serious social scientist to make of all this? These quotes, and the rest of the article, consist largely of claims that lack any kind of empirical backing. Some claims are sociological: that the "social environment" of whites "protects and insulates them from race-based stress." What "social environment"? How measured? Apparently also, whites (All whites? The whites in the room?) "hold . . . institutional power over people of color"? Where is the proof, other than agreement from other students of critical race studies?

Apparently,

> Whiteness scholars define racism as encompassing economic, political, social, and cultural structures, actions, and beliefs [so-called systemic racism] that systematize and perpetuate an unequal distribution of privileges, resources and power between white people and people of color.[50]

What are these "structures, actions, and beliefs"? The concept of "systemic racism" is usually justified (if it is justified at all) by reference to racial disparities—"unequal distribution" of income, health, living conditions, et cetera—most, if not all, of which can be explained in nonracial ways.[51]

DiAngelo refers to words like "urban," "inner city," and "disadvantaged" as "racially coded." Again, where is the proof that the words don't mean just what they say? It's cheating to change the meaning of your critic's words.

DiAngelo's writing is clever rhetoric. She manages to persuade while presenting not a jot of scientific evidence. The writing brims with emotive terms that appeal to the compassion and perhaps guilt of her white audiences.[52] But anecdotes from an arbitrary population, not to mention unsupported claims and allegations, are not science, but propaganda.

Consider that powerful piece from DiAngelo's article I quoted earlier. Here it is again, slightly altered:

> White people in North America live in a social environ-
> ment that subjects them to constant accusations of racism.
> This stressful environment builds white expectations for
> racial antagonism while at the same time lowering their
> ability to get along with other races, leading to what I refer
> to as *White Hostility*. White Hostility is a state in which
> even a minimum amount of racial stress becomes intoler-
> able, triggering a range of defensive moves. These moves
> include the outward display of emotions such as anger,
> fear, and guilt, and behaviors such as argumentation,
> silence, and leaving the stress-inducing situation. These
> behaviors, in turn, function to create racial disharmony.

Is this fragment not also *irrefutable*? Is it any less persuasive than DiAngelo's original? Should we now believe in white hostility rather than white fragility?

In fact, neither white fragility nor white hostility has any factual basis beyond the concurrence of a sympathetic audience. I leave the reader to decide whether DiAngelo's iconic, supposedly academic paper (and the book that followed) is anything more than scientist propaganda for a particularly divisive view of race in America.

Critical Race Theory

Critical Race Theory (CRT) is a movement that has been fermenting in law schools and the humanities and social sciences for thirty years or more. It is not theory in the usual scientific sense: a small number of assumptions from which many testable predictions can be derived. Rather it is a way of racializing almost any social issue. At first it was rather lost in a crowd of other then-fringe movements such as radical feminism. (For example, "our critique of the objective standpoint as male is a critique of science as a specifically male approach to knowledge. With it, we reject male criteria for verification."[53]). But, somehow, the addition of *race* to the intellectual *bouilla-baisse* gave CRT an impetus that has led many scholars and academic administrators to pay attention. A claim such as Kimberlé Crenshaw's that "conceptions such as colorblindness and merit functioned as rhetorics of racial power,"[54] which might once have been dismissed as a toxic racialization of race-neutral ideas is now taken quite seriously.

CRT is a third-stage derivative of Critical Theory, a mid-twentieth century Marxist-related movement advocated by Frankfurt School writers Theodor Adorno, Max Horkheimer, and others. It gave rise to Critical Legal Studies, a 1970s movement applying some of the same ideas to the study of law. CRT now informs much of race and ethnic sociology, not to mention K–12 education.

There is a huge CRT literature, much written in artfully tendentious prose: "recognizes" is regularly used instead of the more honest "assumes" or "claims" (as in "CRT *recognizes* that racism is not a bygone relic of the past [emphasis added]"[55]), and similarly "theorized" instead of "hypothesized," and so forth. But there is a way to evaluate this field and indeed any other that pretends to embrace all human experience in a *scientific* way:

- What are its values?
- What are the facts with which it deals?
- How good is the logic of its arguments?

Applied to standard social science—much of sociology, personality and social psychology—these questions can be simply answered: Values? Just finding the truth, established by the usual scientific method of observation, experimenting where possible, deliberating, and hoping to arrive at a consensus. Facts? Demographic data, results of experiments, randomized samples, third-party-measurable data. Logic? Well, just logic: "If *A* implies *B*, then produce *A* and see if you get *B*," and so on.

CRT is different in every respect. Boiled down to fundamentals—values, facts, and logic—CRT can be ruled out as any kind of science.

Values: "Unlike some academic disciplines, critical race theory contains an activist dimension. It not only tries to understand our social situation, but to change it."[56] It is not science, then, but social engineering, at best. The scientific principles on which such engineering must rest are yet to be discovered, and the aims of CRT are questionable if not racist. The change desired is politico-religious, not scientific, but to "enhance the *power*" of black and other "marginalized" minorities; it is obsessed with "whiteness" as an oppressive norm; it is not, and does not aspire to be, neutral. It firmly rejects any notion of color blindness.

Truth: "Because there are many ways of knowing, there are racial, gendered, and class-centered truths that can only speak for themselves."[57] There is no allegiance to the ideal of consensual truth, a fundamental principle of science. CRT "rejects the prevailing orthodoxy that scholarship should be or could be 'neutral' and 'objective'"[58]—in which case, it qualifies not as scholarship but as propaganda. As for facts, CRT has an "insistence on 'naming our own reality.'"[59] A society's "truth" is often just a way to exert power; truth has no independent existence. Each individual has his own "truth"— something called *standpoint* epistemology. But if you think a porpoise is a fish and I think it is a mammal—or you think I am a racist, but I don't—no resolution is possible. Our standpoints are at a standstill—and science, not to mention harmony—is impossible.

Nevertheless, CRT thinks that some "borrowing of insights from social science on race and racism" is okay.[60] But mainly the emphasis is on "counter-story telling" and personal experience, that is, data not accessible to a third party.

Logic: CRT undervalues logic; the term "racial logics" is sometimes used. Some versions of CRT even consider logic a feature of "whiteness," as in *White Logic, White Methods*.[61]

In short, CRT is neither legitimate social science nor law in the Western tradition. It is not science because it denies the idea of truth and advocates action rather than discovery: it is about *ought* not *is*. It is probably not law because it is more interested in power than justice: it values equality of results ("equity") over equality of treatment. It is an ideological, politico-religious movement, not unscientific so much as a-scientific. It has no credibility in scientific discussions of race. Whether it should be used in K–12 education—a current controversy—depends on the course and the politics: It might find a small place as part of an advanced history course on post-Enlightenment intellectual movements. As a standalone effort, it is indoctrination.

• • •

In 1985, philosopher Roger Scruton penned a *Times* op-ed called "The Plague of Sociology."[62] The problems of sociology have been apparent for many years. How did the field lapse from Durkheim's high standard? Political forces are always present. In the "harder" sciences they are restrained by rigorous methods of experiment and theory that are universally accepted. Sociology began this way, but differences soon led to many divisions, with each new branch accepting a different set of standards for what constituted valid data and acceptable methodology. This separation reduced the variety and force of criticism. Soon, everyone in RES sociology agreed that anecdote is okay, storytelling is as scientific as chemical analysis, my truth may be different from yours, and activism is good scholarship.

The root problem seems to be the endless subdivision of the social sciences. Some means must be found, if not to abolish, at least to mitigate this protective isolation of subdisciplines. Perhaps a modification of a peer-review system that, at present, allows hyper-specialized and often politically involved social-science scholars to listen only to the like-minded. Perhaps social-science grant applications should be vetted by a broader range of scientists, a majority from outside the subspecialty. Perhaps some other solution can be found to restore the academic status of sociology. For the time being, all we can do is highlight what seems to be a widespread and pernicious degradation of science.

CHAPTER 13

Systemic Racism:
What Do Racial Disparities Really Mean?

Racist attitudes of whites toward blacks have become socially unacceptable in America, although the reverse, racism of a minority directed at the white majority, is still tolerated if not encouraged.[1] On the other hand, statistical racial disparities persist. African Americans, as a population, continue to suffer deficits of income, crime and incarceration-rate, health, housing, and family-structure in comparison with the white and Asian populations. These disparities are taken to imply—indeed, are sometimes equated to—*systemic racism*.[2]

The idea of systemic racism originated in the Black Power movement and is still about power. Via academic sociology, it has become a popular explanation for almost every racial ill. The U.S. Conference of Catholic Bishops is a believer, for example: "Today's continuing inequalities in education, housing, employment, wealth, and representation in leadership positions are rooted in our country's shameful history of slavery and systemic racism."[3] President Biden, not to mention the administrators at the National Science Foundation, agrees.[4]

Does this influential concept have a scientific basis?[5] It seems to be defended in two ways: One is empirical, that is, referring to something that can be measured. The other is simply justification by assumption—systemic racism just *is*.

In most writing on the subject, systemic racism is simply assumed. Statements like this, from a *New York Times* article about students' feelings about race, are common: "Students engaged passionately on core issues like the existence of white privilege, the extent of systemic racism, the legacy of slavery . . ."[6] There is some uncertainty about "white privilege," but debate only about the extent of systemic racism, not its existence. Systemic racism is as real as COVID-19 or the national debt.

Disparities = Systemic Racism?

Racial disparities in mortgage lending, income, wealth, housing, crime, and incarceration, et cetera seem to be the only offered empirical evidence of systemic racism. For example, Ashley "Woody" Doane of the University of Hartford claims that "[systemic] racism *is* embedded in the social and political institutions of the United States" using as evidence "the disproportionate impact of mortgage lending upon blacks and Latinos."[7] Others are more sweeping: "As an anti-racist, when I see racial disparities, I see racism,"[8] says Ibram X. Kendi, who directs the Anti-Racist Research and Policy Center at American University.[9] He adds that, "Racial discrimination is the *sole cause* of racial disparities in this country and in the world at large [emphasis added]."[10] A cheery racial-justice website simply equates disparities with systemic racism.[11]

How fair is that? After all, no one thinks that the dominance of women in the nursing profession is proof of sexism or the dominance of blacks in the National Football League a sign of racism. It should be obvious that we need to look beyond disparities—in numbers or wealth or incarceration—beyond disparities to their causes. Thomas Sowell has written several definitive books on the issue, none even mentioned by CRT "scholars."[12]

Disparities will likely persist because individuals differ, and if individuals differ, so will some groups. Under relatively free conditions, and with no prejudice at all, some will rise more than others in any given profession. Hence, given the usual nonrandom distribution of abilities and interests, then even in a society with no discrimination, some disparities, group as well as individual, will always exist (so "anti-racism" à la Kendi has an assured future).

In addition to explaining racial disparities, systemic racism performs another function. It allows the charge of racism to stand even if no individual white person behaves in a racist way. In other words, even if she is genuinely color-blind and follows faithfully Martin Luther King's credo to treat people according to the content of their character and not the color of their skin, even if she is conscientious, decent, and pure in heart, even so she is guilty of color-blind racism. (In a less forgiving—or fearful?—white population, this allegation would surely be regarded as a gratuitous insult.)

Absent evidence of individual racism and given that individuals usually behave in color-blind ways, the supposed existence of a pervasive systemic racism implies *"Racism without Racists."* Bonilla-Silva's provocative book might as well have been called *Racism without End* since, allegedly, disparities will always prove racism exists, and disparities will never vanish—unless, that is, the state enforces a totalitarian "equality of results," which is precisely the solution proposed by many color-blind-racism (CBR) sociologists.[13]

Disparities cannot easily be explained by pointing to racist behavior by whites. The disparities have either increased or remained the same while individual racist behavior has declined. Disparities do exist, but giving them a label—systemic racism—doesn't explain them. We need to inquire as to their cause. There are two possibilities: causes within individuals—what I earlier called *endogenous* (material) causes—or external, *exogenous* (efficient) *causes*.

Endogenous Causes of Racial Disparities

Endogenous causes were in fact the first ones to be studied, with unfortunate results. Bigots stigmatized the entire black "race" as "inferior" because of lower average scores on (for example) IQ tests. Blacks' lower status, health, incomes, and so forth were then comfortably attributed to their built-in inadequacy. But avoiding a question because a half-understood answer has been misused is both unwise and unscientific.

The usual presumption was that IQ is fixed at birth, that it is the most important factor in life success, and that it cannot be altered by later experience. There is some truth to each of these presumptions, but none is the whole story. IQ is indeed a predictor of socio-economic success in Western-type civilization,[14] but it is far from the only factor. IQ is indeed more or less fixed in adulthood, although perhaps less so than was once believed.[15] Fixity of IQ at birth seemed to be supported by several studies showing relatively high (statistical) *heritability* for IQ.[16] But high statistical heritability for a behavioral trait does not imply that it is independent of the rearing environment.

Language is the most obvious counterexample. Statistical heritability just means that children's traits resemble their parents'. Language is 100 percent heritable in this sense: children speak the language of their parents. Language is also 100 percent learned. It is an entirely learned behavior that also has high heritability in this very limited sense. We know that language is not an instinct from "natural experiments" provided by adoption. Despite the high heritability of language, adoptive infants learn the language of their adoptive, not their bio-logical, parents. It follows that high statistical heritability does not mean genetic determinism.

It is worth emphasizing that the role of genes in human ability is less relevant to social policy than the role of the environment. The most important practical issue is the *modifiability* of human traits and abilities, not their genetic components. Genotype is fixed; phenotype—behavior—not so much. It is an undoubted fact that human beings can be ranked on a number of cognitive scales, that this ranking changes little after adulthood, and that persistent inequality will always pose problems for an egalitarian society. It is the apparent rigidity of this ranking, not its supposed genetic basis, that poses an immediate problem for an egalitarian democracy.

On the other hand, real genetic heritability does have an implica-tion for the persistence of social classes. If bright kids do better and, perhaps because of assortative (like-marries-like) mating, their kids are also brighter than average, then a meritocracy will tend toward a hereditary merito-aristocracy.[17] But this is a problem for more research and another day.

Genes and IQ: Trans-Science, Again

What kind of experiment would be needed to prove that intel-ligence, which is statistically heritable (the parent-offspring correla-tion is between 0.6 and 0.8 in the populations studied[18]), is in fact genetically determined? What would it take to establish rigorously the reality of genetic differences in IQ between races? A lot, it turns out: the question is another example of Weinberg's trans-science. Only a very elaborate, and in practice undoable, set of experiments would suffice.

We lack now and for the foreseeable future detailed understanding of how the human genotype produces the human brain, although baby steps are being made to unravel the puzzle.[19] Nor do we know exactly how the brain works to produce human mind and behavior. We cannot therefore map out step-by-step, in detail, how genes-build-the-brain-makes-behavior. Therefore, to understand the genetic determination of IQ, we would be forced to resort to experiments with whole human infants. To address the black-white issue in a completely rigorous way, we would need to begin with two fetal genotypes, one "black" and one "white." Next, we would need an indefinite number of identical copies—clones—of each genotype. Then, each clone would be exposed to a different rearing environment. (Have we tried out the full range—all relevant environments? How would we know?) Finally, perhaps sixteen years later, we would give IQ tests to each child. We would get a distribution of IQs for each genotype. We could then ask: How do these distributions differ? Are their means the same or different? Which is higher? Is one distribution more variable than the other?

Suppose we find a mean difference? Does that settle issue? Well, no, not yet. We are talking about *race* differences here, not differences between individual genotypes. A race is a population, a range of genotypes. So, we need to repeat this impossible process with an indefinitely large sample of "white" and "black" genotypes. (There will, of course, be debate about which genotypes go into which group.) Only after we have this two-dimensional array of genotypes versus IQ can we compare them and come up with a valid conclusion about race difference in IQ.

An elaborate study of this sort is of course impossible. But what about adoption? Adoptive infants learn the language of their adoptive parents; do adopted children acquire the IQ of their adoptive parents? There have been several studies along these lines. A landmark 1997 study looked at 245 adoptions within a mostly white population. The conclusion of Plomin and his fellow researchers was unequivocal:

> Children increasingly resemble their parents in cognitive abilities from infancy through adolescence. Results obtained from a 20-year longitudinal adoption study of 245 adopted children and their biological and adoptive parents, as well as 245 matched nonadoptive (control) parents and offspring, show that this increasing resemblance is due to genetic factors. Adopted children resemble their adoptive parents slightly in early childhood but not at all in middle childhood or adolescence.[20]

The convergence of IQ between biological parents and children with age is also true of siblings, who grow more similar to their parents in many psychological measures well into adulthood, as the (statistical) heritability of IQ increases substantially with age.[21] These results are discouraging for advocates of "nurture" versus "nature" as the exclusive source of racial IQ differences.

The study did not compare white and black kids. It is quite possible, therefore, that black kids will show less influence of parental IQ than white kids do. In which case transracial adoptions, black kids adopted by white parents, might well show more convergence on their adoptive parents than Plomin's white kids did. In a carefully worded sentence, the Minnesota Transracial Adoption Study follow-up concludes: "putative genetic racial differences do not account for a major portion of the IQ performance difference between racial groups."[22] Or, more directly stated: racial differences account for less than half the black-white IQ difference. Clearly the role of genetics in IQ differences is not negligible, as it seems to be for language learning, but neither is it the dominant factor.

We don't understand how genes affect the development of the individual brain and we cannot do the kind of experiments necessary to get a definitive answer to the genetic black-white-IQ-difference question. Adoption "natural experiments" are necessarily flawed: neither adoptees nor adoptive parents are randomly selected, and age at adoption cannot be controlled (some children are adopted as infants, others as pre-teens or older). The impossibility of scientific proof has left the question wide open to other influences, as we will see.

The Equality Imperative

[A] huge swath of Americans . . . believe that black people are not as smart as white people. Or believe black people are somehow more criminally inclined.
—Ta Nehisi Coates

No, we're never running s**t like that, obviously.
—Jeffrey Goldberg[23]

People differ in their abilities, and after a certain age these differences are more or less fixed. Received wisdom in the United States has in recent years increasingly downplayed these facts. It has been especially hostile to the possibility that individual differences are not randomly distributed across, for example, the two sexes or different income or racial groups.

Practices that make individual differences hard to ignore are condemned. There are moves to abolish grades, for example, called "relics from a less enlightened age" by one education expert.[24] A 2020 *New York Times* article subtitled "What's the Point of Grades?" (written by an "assistant professor of leadership education"[25]) also concludes that grades should be abolished. The article quotes a Wellesley college official who says that pass/fail is to be adopted so as to "support one another without being required to make judgments."[26] Apparently, Wellesley professors find it painful to make judgements about their students' performances. Judgment is tough for the judge (all that grading) and sometimes vexes the judged.

In the decades before, and for a while after, the Second World War in the United Kingdom and the United States, grades in high school and college were made public with no apology: seeing, and revealing, how you did in relation to others was regarded as a useful incentive to effort. A professor recounts how "tracking" operated in the 1960s. In his California school, even seating was arranged according to grade point average, updated after every test. Everyone knew who was in the top row.

With time, broadcasting grades fell out of favor everywhere, rationalized by claiming that grading interferes with "real education."[27] Concerns were also raised about students' "mental states." Being seen as low down on the grade list made some students unhappy and perhaps an object of scorn to their classmates. Making grades common knowledge is now regarded as deplorably stressful as well as embarrassing.

There are other reasons for the flight from grades. In 2019 Yale law professor Daniel Markovits published a widely reviewed book titled *The Meritocracy Trap: How America's Foundational Myth Feeds Inequality, Dismantles the Middle Class, and Devours the Elite.*[28] The book's first sentence is "Merit is a sham," the last chapter is "The Myth of Merit." Convinced that "Meritocracy, like aristocracy, comprehensively isolates an elite caste from the rest of society and enables this caste to pass its advantage down through the generations," Markovits seems to object to any kind of social hierarchy, even Thomas Jefferson's "natural aristocracy" based not on "wealth and birth" but rather "virtue and talent."[29]

Not to be outdone, "TED" star and Harvard law professor Michael Sandel, named "most influential foreign figure of the year"—in China[30]—published *The Tyranny of Merit* in 2020.[31] Sandel's main beef is the arrogant attitude of the successful: these kids think too much of themselves and too little of those who failed to achieve as they have.

Sandel makes three claims:

- That society is stratified by access to elite colleges and universities: a graduate from an elite college does way better in life than someone without a Harvard degree or indeed without any degree[32]
- That stratification by merit is unfair, even if merit is objectively assessed, partly because the successful tend to come from rich families
- That the merit system makes arrogant those who succeed and humiliates those who fail[33]

On the other hand, paradoxically, Sandel concedes, that most "agree . . . that admission should be based on merit."[34]

First, as to the arrogance of the successful and the resentment of the rest: This is obviously not a *necessary* byproduct of a meritocracy. These attitudes reflect the attitude of society and its effects on meritocratic winners. If all regard the criteria for success as basically fair, the successful need not be arrogant nor the unsuccessful resentful. In sport, for example, rank is celebrated not deplored. Sandel's problem, if there is a problem, is society and the way merit is assessed, not meritocracy per se.

Sandel is upset by what he sees as the unfairness of the merit system. He sees it as unfair in two slightly contradictory ways: (a) because successful kids tend to come from successful families, implying that "merit" is more an index of wealth than talent; and (b) and because even if merit is accurately assessed, why do the talented deserve more?

The first objection seems to imply that the success of upper-income kids at standardized tests is somehow attributable not to their talent and efforts but to coaching which is unavailable to poorer kids. This (like everything in this area) is a controversial issue, but these tests are designed explicitly to measure the kind of talent that cannot be effectively taught. Despite one or two books making the case for malleability,[35] it is likely that adult IQ cannot be shifted much.[36] Coaching enterprises proliferate,[37] but their success probably owes more to demand than outcome. There is also considerable evidence that ability is partly inherited, so some of the benefit of being wellborn is . . . being wellborn.

Sandel's second objection is more problematic: What is "fair"' about being born with exceptional talent?

> Is having (or lacking) certain talents really our own doing? If not, it is hard to see why those who rise thanks to their talents deserve greater rewards than those who may be equally hardworking but less endowed with the gifts a market society happens to prize.[38]

Indeed, there is nothing fair about what Sandel calls the "genetic lottery"—that Aishwarya Rai is beautiful and Richard Feynman is smart, not to mention a random sample who are neither. On the other hand, unless you are a Marxist or a Buddhist, reward should bear some relation to skill and effort; and incentives related to value produced seem to be necessary to a successful society. A society that seeks to rectify "genetic injustice" is proposing to create the social equivalent of a perpetual motion machine.[39] It is a truly intractable issue, a matter for politics, religion, and moral philosophy, not science. It is not best solved by reducing everyone to the lowest level.

But unhappiness with meritocracy leads to a fallacy that *is* a scientific issue. In a telling aside, Sandel comments:

> Success at making money has little to do with native intelligence, *if such a thing exists* [emphasis added].[40]

A contemporary myth is that there is no such thing as people who are "smart" as opposed to "dumb," or even "able" as opposed to "not so able."[41] Best sellers like Malcolm Gladwell's *Outliers* claim that persistent hard work—the "10,000-hour rule"—will propel anyone to success, with the implication that there is no such thing as natural talent.[42] We could all compose like Mozart if we just practiced long enough—which is wishful nonsense.[43] There are people who are smart and people who are not, and there are different kinds of "smart." Similarly for "drive," "guts," and many other valuable attributes. Society needs to recognize the obvious, not try to hide it.

The playing field of the nature-nurture debate has not been, and is not now, level. Even the most rational and disinterested writers on this topic have been attacked and their employment threatened on account of purely academic, nonideological writing. A typical victim is philosopher Michael Levin, who was criticized in hostile and dishonest ways following the publication of his thoughtful 1997 book *Why Race Matters*.[44]

The view that IQ is genetically fixed—like an instinct—is in fact held by none of the leading researchers in this field. Yet the charge has been used to defame honest scholars[45] such as Linda Gottfredsen[46] and Charles Murray.[47] Well-known opinion writers concocted incendiary links between coercive eugenics laws and anyone who dared to discuss the heritability of IQ (not to mention hints at "white supremacy" and links to race-based Nazi genocide).[48] Data and arguments were buried under a hailstorm of ad hominem attacks.

Fear of treatment like this from the Southern Poverty Law Center (SPLC) and similar groups, eager to stigmatize,[49] but with no interest in understanding the complex issues involved, has in effect shut down research on endogenous differences as one cause of racial and other disparities. We still do not understand the causes of individual differences in IQ and other racially nonrandom cognitive and personality characteristics. What is the role of rearing environment, culture, and social context? Of genetic endowment? How do these individual differences interact with the social environment to bring about the disparities that so dismay many social scientists? Even severe critics such as Charles Lane concede that "blacks consistently have scored lower than whites on IQ tests,"[50] but research has yet to establish either the reasons for this IQ difference or its role in real-world black-white disparities. Nor will it, given the rancor and disinformation that surrounds these questions; nor will we know if these differences persist or change over generations. A visiting extra-terrestrial, noticing the asymmetry of the political climate, might well conclude that the violence of these attacks suggests that there must be something to an extreme hereditarian position!

Exogenous Causes of Racial Disparities

With endogenous causes off the table, all that is left to account for racial disparities is exogenous causes. Racism, systemic and otherwise, is the favored cause, but less-favored environmental factors, like schooling, home, and neighborhood culture, are probably much more important.[51] Almost everyone admits that discrimination by individuals has decreased in recent decades. A University of Illinois survey concluded that "One of the most substantial changes in white racial attitudes has been the movement from very substantial opposition to the principle of racial equality to one of almost universal support."[52]

The absence of discriminatory attitudes doesn't mean the playing field has been completely leveled, of course. A past bias may persist long after the world has changed. Robert Moses's racist-inspired building practices in New York leave their mark today.[53] It's not hard to point to lingering effects of past racism on parents, children, and neighborhoods. These effects act in complex ways. Proving their existence, much less measuring them with precision, is almost impossible. Many effects are delayed, different individuals vary in their reactions, and allegedly discriminatory actions are often unintentional or a byproduct of nonracial factors.

The problem is that these complex and little-understood effects, plus still imperfectly understood endogenous differences, have been bound together into the toxic bundle of systemic racism.

Individual and Systemic Racism Defined

The word *racism* is thrown around a lot, but a precise definition is hard to find. *The Stanford Encyclopedia of Philosophy* passes on a definition. In 2020 Wikipedia came up with "**Racism** is the belief in the superiority of one race over another . . ."[54] (The current definition is more tendentious.[55]) Other definitions add reasons for racism; particularly popular is the idea that *inferior* races have unchangeable (*genetically determined*) attributes that render them less able than the favored (usually *white*) race. But a cause is unnecessary. Basically, racism is intentional discrimination against a race for no reason at all: you just value white people over black people (or vice versa).

How about *systemic*—also known as institutional or structural—racism? How is it to be measured? We can get a hint from the British Macpherson Report, produced by an official commission in 1999 as a reaction to the racially motivated death of a young black boy.[56] The report is clear about individual racism, which "consists of conduct or words or practices which disadvantage or advantage people because of their colour, culture, or ethnic origin."[57] But Macpherson repeatedly hesitates to define institutional racism, finally concluding that it is:

> the collective failure *of an organization* to provide an appropriate and professional service to people *because of their colour*, culture, or ethnic origin [emphasis added].[58]

That seems clear enough: Institutional racism is when people are treated differently by an organization or institution because of their color. Racial disparities by themselves are not enough. The fact that African Americans are underrepresented among professors of, say, physics, does not by itself demonstrate racism, systemic or otherwise.

By none of these definitions can the complex, lagged effects of past discrimination be termed racist. The effects I will discuss are not intentional—if they were, they would probably qualify as individual, not systemic racism. They are not based directly on race, for the most part. If a black child fails to get into a good college because she scored poorly on the SAT, she fails because of her score, not her race—even if her low score might be traced to poor rearing conditions that are a legacy of segregation or past discrimination. Effects like this are, possibly, effects of racism in the past. They are not racism now, systemic or otherwise, by any reasonable definition. Failure to acknowledge this distinction has unjustly stigmatized white people today and is a cause of needless conflict and misdirected effort. Here are more examples of the problems with this insidious concept.

More Disparities: That Elusive Systemic Racism

A 2017 review paper in the respected medical journal *The Lancet*, authored by several public-health researchers, looks at the health implications of what the authors term *structural racism*. They refer to the "rich social science literature conceptualizing structural racism" emphasizing that the idea goes beyond "unfair treatment as experienced by individuals." Yet in the next paragraph they say "Any account of structural racism within the USA must start with the experiences of black people . . ."[59] This inconsistency is never resolved: Is "black experience" relevant, or is it not? Apparently not, as the paper goes on to discuss racial disparities as evidence of structural racism.

Crime and Punishment

One example is this: "The legacy of these [ostensibly race-neutral] policies is that the annual rate of incarceration of black men is 3.8–10.5 times greater than that of white men, across all age groups," which is obviously unfair, hence racist.[60] Unfair—*unless the rate of offending is also skewed*: Do blacks in fact commit proportionately more crimes than whites?

The SPLC, which sees racism everywhere, has weighed in on this issue, citing a relevant report from the U.S. Bureau of Justice: *Race and Hispanic Origin of Victims and Offenders, 2012–15*. They headline their article:

> White Supremacists' Favorite Myths about Black Crime Rates Take Another Hit from BJS Study: Vast Majority of Most Crimes Are Committed by a Person of the Same Race as the Victim, Bureau of Justice Statistics Reports.[61]

They go on to say that "White supremacists . . . claim that . . . African-Americans, are far more crime-prone and the source of most violent crime against whites."[62]

The headline misses the main point of the BJS report. First, the U.S. population is 65 percent white and 12 percent black. It is likely, therefore, that more crimes can be attributed to whites than to blacks. No surprise there. And it has been known for many years that most violent crimes are intra- not inter-racial: blacks are the victims largely of black criminals, whites of white.

The real issue is "crime-proneness," which depends both on the number of black and white perpetrators and on the sizes of the black and white populations. Population size is not mentioned in the SPLC article. There are fewer blacks than whites in the United States, therefore we can expect fewer black than white perpetrators. But how many fewer? Well, if we include population figures (which do appear in the BJS report), we see that there are 5.3 times as many whites as blacks in the United States. So, the real question is, are there 5.3 times as many white as black criminals? Well, no: from table 1 in the BJS report, 43.8 percent of perpetrators are white and 22.7 percent are black, so the ratio of white to black criminals is just 1.93. In other words, blacks are 5.3/1.93 = 2.75 times as likely to be perpetrators as whites.[63] Blacks are indeed *more* violent-crime-prone than whites, contrary to the SPLC report. It is hard to know whether the reporter is guilty of willful distortion or simple incompetence: this is an elementary error.

Again, these data need to be unpacked to understand what is really going on. Young males are more likely to act violently than older ones. The black population tends to be younger than the white. Does controlling for age reduce the black-white disparity? No doubt other relevant variables (poverty, family structure, local gang culture, and so on) should be examined. The point is that the incarceration disparity may have a nonracial cause. Absent a lot more research it should not be blamed immediately on racism.

The BJS report deals with all violent crime, a measure that is imprecise for several reasons: the definition of "violent" varies from one person to another, and there may be differences in reporting between black and white communities—blacks may be more (or less) reluctant to report crimes than whites, for example. *Murder* statistics are necessarily more certain than reports of nonlethal crimes. Here, the data are unequivocal. The Centers for Disease Control (CDC) in a 2017 report concludes: "The homicide rate among African-Americans is nearly quadruple that of the national average" and some *eight times* the rate for whites.[64]

So, there is some nonracial basis for the apparently racist fact that blacks are incarcerated at a higher proportion than whites. Blacks are also more likely to commit violent crimes, and much more likely to commit murder, than whites. Of course, the details of the causality—in particular, the possible links between black criminality and racism suffered in the past—are not at all understood. But what should be understood is that this disparity is no proof of contemporary racism. Much the same argument applies to racial profiling—racial disparities in police "stop-and-search," for example.[65]

Incarceration rate is a clear case where a racial disparity can be traced to a nonracial cause rather than racism, systemic or otherwise. There may also be racial causes; but they cannot be accurately assessed so long as nonracial causes are ignored.[66]

Gerrymandering, et Cetera

The *Lancet* article goes on to list other examples of systemic/structural racism; how valid are they?

The Social Security Act of 1935 is a particularly poignant object of their scorn. The authors attribute the act to an attempt "to secure the votes of Democrats in the South."[67] Redistricting to restrict black voter participation, "racial gerrymandering," is cited as another example of systemic racism.[68] But in both these cases, the aim is to secure votes, not to exclude black people because they are black. If the African-American population changed its preferences from Democrat to Republican, no doubt gerrymandering practices would adapt accordingly. The point is that these practices are not aimed at black people because they are black, but because of the way they vote. Gerrymandering is sleazy, but it is not racist—systemic or otherwise.

Likewise for so-called *voter suppression*, a term applied to practices such as implementations of strict voter-identification laws, closings of department of motor vehicle offices, restrictions of early voting, et cetera. For the most part these practices are deemed "racist" not because they actually discriminate by race, but because (it is presumed—detailed evidence is rarely offered) a larger proportion of black people than white people are adversely affect by them. The remedy is either to fix the identification problem or live with a color-blind disparity, not distort the voting system in a way that would be (reversely) racist. So long as voting is not actually tied to race, the fact that one race may in some places be disadvantaged vis a vis others does not make the practice racist. Once again, racial disparities do not equate to racism, systemic or otherwise.

Systems Theory

This kind of muddle infuses the *Lancet* article and many others like it. But the fact remains that there are certainly lingering effects of past racism. How should they be understood? Labeling them as systemic racism is not helpful, because it suggests a single cause for what are in fact multiple interacting causes, most of them nonracial. One solution has been to invoke systems theory as a way to model a complex situation. A *system* is "an organized entity made up of interrelated and interdependent parts." The basic idea is that no part can be considered in isolation because all are connected. An action on one part will have effects on many others. Systems thinking has been successfully applied to collective phenomena such as flocking in birds and schooling in fish, explaining coordinated behavior by leaderless groups in terms of simple rules followed by each individual group member. In this case the emergent, coordinated behavior can be measured precisely, and hypothetical rules can be rigorously tested by computer simulation.

But systems theory has been invoked much more widely, in areas from political economy to social work. The social-science version of systems theory cannot be accused of excessive precision. A sociology paper begins with this: "An example of an emergent property is wetness. Neither hydrogen nor oxygen alone has or can produce wetness; wetness occurs only in a chemical system that includes both hydrogen and oxygen in the correct proportions."[69] The problem is that the sociological equivalents of the elements hydrogen and oxygen can neither be accurately identified nor independently manipulated.

A paper that seems to have influenced contemporary thinking in sociology is Thomas Schelling's 1971 "Dynamic Models of Segregation," which has as one conclusion: "In some cases small incentives, almost imperceptible differentials, can lead to strikingly polarized results. Gresham's law is a good example."[70] This is an early version of the "tipping point," the idea popularized by Malcolm Gladwell to explain how small trends take over whole societies. This idea has appeal to students of racism, since it allows the possibility that even small vestiges of racism may be amplified by "race-discrimination-system" interactions to produce what Barbara Reskin has called "über racism."[71] It's certainly possible: chaos theory has given us the *butterfly effect*, after all. But large effects from small causes, or even from large but mostly unrelated, nonracial causes, are the exception rather than rule. Über racism requires much more convincing proof than the kind of qualitative, even impressionistic, accounts offered by workers in this area.

There are two other problems with this approach. The first afflicts all of sociology. There is much discussion of "causal models." But since experiment is generally impossible, only *correlations* remain. If these correlations are sufficiently compelling,[72] perhaps causality can be inferred, but in fact they rarely are. The second problem is that in the racism area there are almost no quantitative correlations at all. What are offered are not so much models as visual metaphors. Researchers paint different, plausible, but completely untestable (hence unscientific) pictures of how historical racism and its contemporary aftereffects might work together to produce disparate racial impacts. They are metaphors. Yet they are used to justify the concept of systemic racism.

• • •

Systemic racism is a bad idea. First, it is almost impossible to prove because racism is discrimination without any reason other than race. To prove discrimination, all other possible reasons—reasons like differential ability, interests, criminality, et cetera, as in the examples I gave earlier—all other possible reasons must be eliminated. Does the tech industry discriminate against women? Does the nursing profession discriminate against men? Or are men more interested in tech and women more interested in nursing? It's not rocket science. To show racism, which is differential treatment for no other reason than race, other reasons for disparities must be eliminated. But in practice not only are they not eliminated, but efforts to explore these other causes are actively suppressed.

The second problem follows from the first: a focus on systemic racism deflects attention from the causes, endogenous as well as exogenous, of the racial disparities that led to its invention. Disparities—racial, ethnic, or gender-based—are not proof of anything.[73] Disparities raise questions about their cause. Absent further information, a racial disparity does not favor one answer over others. To say, as Ibram Kendi and others have, that "When I See Racial Disparities, I See Racism" is either ignorant or irresponsible.

Disparities—in mortgage lending, wealth, or incarceration—can be explained in several ways, of which systemic racism is only one. In other words, racial disparities pose questions. They do not provide an answer. Answers could be found. But the taboos against researching other potential causes of racial disparities—family structure, the abilities and interests of African Americans versus whites, et cetera—have turned out to be almost insurmountable.[74] The consequences of enforced ignorance may be dire. Instead of a scientific answer, "systemic racism" provides only a malign mirage that constantly retreats as we approach. The fact that so many public figures, in business, politics, and academe, pay homage to this baseless and destructive concept raises troubling questions, not about racism but about their own fitness for office.

Systemic racism has become the elusive and inexpugnable cause of all the ills of people of color. It implies racism without end, a virus for which there is no vaccine. And it provides an endless supply of ammunition for those whose careers depend on the persistence of racial anxiety. It has become a cause of racial division rather than part of the cure. It is not science; it should be abandoned.

Chapter 14

Lysenko Redivivus

"I know of no country in which there is so little independence of mind and freedom of discussion as in America."

—*Alexis de Tocqueville*

Suppressing a line of research can be very dangerous. In the 1930s Trofim Lysenko, a dour fellow of peasant origin, acquired considerable influence over the study of genetics in the Soviet Union. As a student, he was advanced by the Soviet version of affirmative action—people of proletarian origin were favored over the "bourgeois," and "Trofim Lysenko satisfied the new criteria perfectly."[1] By 1940 he had become dominant as director of the Institute of Genetics and protégé of Marxist-Leninist dictator Josef Stalin.

Marxist theory is not evenhanded in the nature-nurture debate: *environment rules*. Lysenko's dismissal of Mendelian genetics and embrace of a sort of Lamarckian view of the inheritance of acquired—environmentally induced—characters well fit the prevailing ideology. His ideas were vigorously enforced, and dissenters were actively suppressed—fired, exiled, even executed. A death sentence was passed on respected and once-favored geneticist Nicolai Vavilov, who had actually supported Lysenko early in his career. It was subsequently commuted, and he died in prison in 1943. (Vavilov has been posthumously revived and is now a "hero of science" with his picture on a postage stamp.[2]) Eminent scientists in the West, such as the Brits J. B. S. Haldane[3] and J. D. Bernal,[4] allowed their allegiance to Marxism to shut their eyes to Stalin's brutality. Despite their own expertise, they hugely downplayed—ignored—Lysenko's deadly error.

The suppression of real genetics was costly not just to Russian geneticists but to the Russian people. "Trofim Lysenko's spurious research prolonged famines that killed millions" as his enforced and ineffective policy of "vernalization" caused crop failures and mass starvation. "Although it's impossible to say for sure, Trofim Lysenko probably killed more human beings than any individual scientist in history."[5]

Social Pressure

The authority of a king is purely physical, and it controls the actions of the subject without subduing his private will; but the majority possesses a power which is physical and moral at the same time; it acts upon the will as well as upon the actions of men, and it represses not only all contest, but all controversy.

—Alexis de Tocqueville[6]

The most effective way to suppress inquiry is not violence but self-censorship, enforced by social pressure. In the contemporary West, research in certain directions is blocked not by the brutal methods of a totalitarian dictator urged on by a wrongheaded scientist, but by an ideological consensus that has been growing since the 1960s.[7]

Petitions, facilitated now by social media, are a popular way to enforce the preferences of an activist group. Here is one example, from many. In March 2021, *The Times* headlined, "Academics Led the Campaign to Silence Genetics Professor Gregory Clark on Race."[8] Clark, a Scottish-born visiting professor from the University of California at Davis, was to give a talk at Glasgow University's Adam Smith Business School. (Clark's talk was in effect an updated version of Francis Galton's 1869 book *Hereditary Genius*.) Clark tentatively entitled it "For Whom the Bell Curve Tolls: A Lineage of 400,000 Individuals 1750–2020 Shows Genetics Determines Most Social Outcomes."

The "Bell Curve" allusion was a trigger for both students and faculty, leading to a collective letter which concluded:

> *We firmly oppose the circulation of genetically-determined ideas of any aspect of social worlds.* Not only because those ideas have been overwhelmingly discredited by reputable and rigorous research, but because due to their history, their circulation as legitimate science has political implications that pose a danger to the mission of the university and the wellbeing of its community [emphasis added].[9]

So, there is no role for genetics in the "social world," not only because it has been "overwhelmingly discredited" (there is in fact *no* research that proves genetics is irrelevant to social issues), but because its study has bad effects. Refutation is not enough: *elimination* is what's required. Lysenko lives in Scotland!

Almost any social-science research on human heredity can provoke opposition. *Paleogenetics*, the study of gene relationships revealed by ancient human remains, has exploded in recent decades.[10] It has allowed the mapping of human migrations over the millennia, and in so doing has challenged a few of the "myths of origin" believed both by those belonging to European Judeo-Christian traditions and North American "First Nations."

Paleogeneticists have celebrated the liberating effect of this work on our understanding of Europeans' origins: "We can falsify this notion that anyone is pure," says population geneticist Lynn Jorde of the University of Utah in Salt Lake City. Instead, almost all modern humans "have this incredibly complex history of mixing and mating and migration."[11]

But these revelations have not been received so warmly by some Native American groups, who want the work to cease. Surprisingly, many scientists have acquiesced to the suppression of their own research, apparently valuing the cultural traditions of some indigenous groups over their own Enlightenment scientific traditions: "Anticipating and addressing the social implications of scientific work is a fundamental responsibility of all scientists," writes a group formed by the American Society of Human Genetics.[12] This seemingly benign provision has had less benign consequences, as we will see.

In 1990, the opaquely written Native American Graves Protection and Repatriation Act (NAGPRA) was passed, requiring any project getting federal funding to return "cultural items" (including human remains) to the "Indian tribe or Native Hawaiian organization" that agrees to accept them. Reaction to this law, which most scientists favored at the time, has now broadened to the point that the preferences of almost any indigenous group are likely to prevail over the needs of research and the values of science, "as the demands of Indigenous communities typically are seen as more morally compelling (especially to academics) than those of the researchers who populate this field of scientific inquiry."[13]

The religious beliefs of Native Americans are not written down. Just like the secularist "faiths" I described in the beginning of this book, they cannot be traced to formal doctrine. The secularists claim (wrongly) to derive their beliefs from reason. Native Americans justify theirs by reference to oral traditions and mythical stories.

Nevertheless, their "oppressed" faith has prevailed over the supposedly dominant Western faith in the value of science.[14]

This story is still unfolding, but two things are clear. First, European "myths of origin" have proved to be no impediment to the scientific investigation of European paleo-ancestry. But comparable myths of Native Americans and Canadian "First Nations" are proving an obstacle to comparable studies of North American prehistory.

The reasons for this disparity are familiar. One factor seems to be the kind of collective "race guilt" which led to national tolerance, even sympathy, for the Black Lives Matter riots of 2020. Another is what philosopher Roger Scruton has called "oikophobia," an embarrassed disapproval of Western civilization.[15] Still another is the neo-Christian "last-shall-be-first" rule, which favors any group perceived as "victim." Whatever the reasons, deference to "marginalized groups" is endemic among the intellectual classes, many scientists included. Consequently, if "marginals" object, the Enlightenment tradition is likely to give way.

A New Norm

A profound effect of these social forces is a change in norms, in those things we take for granted. Consider, for example, a recent *Proceedings of the National Academy of Sciences* article by three Princeton University social scientists, a tiny sliver of a voluminous literature. These scientists studied the promotion of racial diversity in colleges and universities. Their article is entitled: "How University Diversity Rationales Inform Student Preferences and Outcomes."[16] The study (actually eight separate studies, one with "caregivers," and two looking at websites[17]) presented hypothetical diversity statements to several hundred online white and black participants and solicited reactions.[18]

Proclaiming a commitment to diversity is now the rule in American universities. The point of the article is to look at how college administrations defend this commitment—and the effect this has on minority students. The article's authors, led by Jordan Starck, ask which kind of diversity rationale is best for what they term "low-status racial minorities," that is, African Americans.

For legal and possibly ethical reasons, most universities defend their commitment to diversity on educational grounds. Harvard University justifies its position this way, for example:

> Harvard's commitment to diversity in all forms is rooted in our fundamental belief that engaging with unfamiliar ideas, perspectives, cultures, and people creates the conditions for dramatic and meaningful growth.[19]

Presumably "meaningful growth" is just edu-speak for "learning" or "education." Harvard's piece is what Starck and fellow authors call an "instrumental" justification for diversity: diversity is good because it makes kids better educated.

Stanford University touches three bases:

1. **Diversity is important to our research and educational missions.** "Our diversity ensures our strength as an intellectual community. In today's world, diversity represents the **key** to excellence and achievement."
2. **The future is diverse.** "We believe that Stanford's future preeminence requires that we enthusiastically embrace our diverse future **now**."
3. **Social justice.** "We **must** continue to evolve and become a better and more inclusive institution in our pursuit of the values we hold dear."[20]

So, an "educational/political" ("in today's world . . .") rationale is combined with an "arc of history," but also a nod to social justice—mostly "instrumental," but partly what the authors call a "moral" justification.[21]

Starck and his colleagues question educational/instrumental-type rationales, not because they are inappropriate or dishonest, but because of their effect on racial disparities:

> We expect that instrumental commitments to diversity engender organizational cultures that *less effectively protect against racially disparate outcomes* [emphasis added].[22]

This seemingly innocuous statement embraces two errors: First, it assumes a causal relationship between an "organizational culture," represented by a diversity statement, and educational outcomes, something a single-variable correlational study like this can never prove. And second, and more alarmingly, it assumes that there cannot, must not be, *racially disparate outcomes*; apparently any racial disparity implies that there is something wrong with an institution. It is the equality imperative made visible.

Despite its mass of statistics and large number of subjects (also known as "participants"), this article is as close to meaningless as you are likely to find in a prestigious journal. It implies causation ("engender . . . disparate outcomes") when the data are only correlations. It has not shown, cannot show, any cause-effect relation between the type of diversity statement and students' graduation rates. It also ignores factors certain to be much more important to graduation-rate differences than "diversity rationales": the difficulty and types (STEM or other) of courses offered by institutions; who takes what courses; and the caliber of the different student groups (the target group and the competition); that is, endogenous and other exogenous causes of academic performance. Graduation rate certainly depends on these things, which may also show correlations with type of diversity rationale, allowing plenty of "p-hacking," so the whole project is an inconclusive mess, ideologically conformist, but scientifically useless. The National Academy of Sciences should be embarrassed to publish such a meaningless piece of work.

Most alarming of all, the article assumes that any racial disparities are simply inadmissible. Equality of results is the new norm.

Instability

A predisposition to groupthink is longstanding: Americans' distressing tendency to "get along by going along," a strange reluctance to debate even, perhaps especially, in the academy seems to be part of the American character.[23] We are apparently still as Tocqueville found us in 1835, willing to censor and punish others by social means and acquiesce to new norms unsupported by reason.

It is also possible that the "science system" itself is unstable in ways that have allowed it to fall into the present dysfunction. In previous eras, there were many fewer scientists and support for science was widely distributed—diverse. Darwin was independently wealthy, Wallace got support from his work as a specimen collector, and Michael Faraday worked for The Royal Institution, a private charitable foundation. In the United States, universities funded much of their own faculty research. Who got the money was decided in different ways in different places, often ad hominem and quite informally: "He's a good man, we should support him . . ."

Since World War II, funding for science has become highly centralized and highly bureaucratized. Most social and biomedical science funding in the United States now comes from the National Institutes of Health and the National Science Foundation. Proposals are vetted by committees of peer experts. In the early days at NIH, a substantial number of awards went to able individuals for relatively long periods: five, ten years, even for life, at first following older practice and allowing freedom to fail. Now most awards are proposal-by-proposal rather than to an individual, usually for five years or less: failure is not so acceptable.

Funding is highly competitive, and the probability of payoff for a given proposal is often well below 50 percent. Mild dissent, or even an absence of enthusiasm from the one or two members of a reviewing committee who specialize in the topic of a given proposal, can doom it. This has surely created an environment where scientists fear to cause offense. If a small number of peers become ideologically infected, while the rest are more or less indifferent, the system, now hypersensitive to peer pressure, allows them to exert a disproportionate influence. Everyone will be afraid to oppose what is seen as a consensus view. It is quite possible that in search of "accountability" and "objective review" we have created a highly unstable system where a small group of ideologues are able to dominate, creating a "cancel culture" that can easily suppress some lines of research and warp others.

Science, like evolution, requires variation—diversity of ideas—to advance. The present science system creates social pressures that push toward consensus ahead of truth.

Whatever the mechanism, there is little doubt that research on individual and group differences, particularly cognitive differences, has been almost completely suppressed, for reasons not wildly different than those in the Lysenko case: belief in the hegemony of the environment.[24] Some individual differences are inherited and relatively immune to environmental influence: practice will not make you a Mozart or me a mathematical genius. But since the 1960s, the United States has an imperative almost as powerful as the Soviet

Marxist one: the imperative of equality, not just legal equality, but equality in every dimension. If people differ, it must be because of their environment, not their genes or anything else that involves only them. But to use this belief to suppress research on endogenous differences is as dangerous to us as Lysenko was to Soviet Russia.

Rationalizing Censorship

Nevertheless, the effort has extended even to philosophy, although usually well concealed by obscure language and even the occasional lapse into probability algebra. For example, more than twenty years ago distinguished philosopher Philip Kitcher justified an effort to suppress research into the causes of racial disparities by reference to a sort of societal feedback mechanism.

He begins by referring to popular biologist Stephen Jay Gould who, in his book critical of IQ-testing, *The Mismeasure of Man*, says how important it is to hold "some human sciences to high standards of evidence."[25] Kitcher first imagines what he calls an "anti-egalitarian" hypothesis, for example, one that favors endogenous racial differences as a cause for a racial disparities. Suppose it equally likely to be true or false, $p = .5$. Then, further suppose that because of pre-existing societal bias, a study that confirms its chances at no better than 50:50 will be taken by an ignorant/bigoted society as sure confirmation. This could harm the affected group and put them in jeopardy.[26]

Kitcher is cautious about this line of argument, but he does accept the idea that the anticipated societal ill-effect of a certain line of research justifies efforts to suppress it. That is, suppressing science can be justified on the basis of a necessarily inexact assessment of social bias. This has repercussions that strike the heart of the ideal of free inquiry, and Kitcher doesn't hesitate to spell those out: "But we ought at least to be aware that the worthy ideal of free inquiry presupposes the absence of conditions that are, unfortunately, found in actuality [emphasis added]."[27]

This is an extraordinary lapse, both ethically and scientifically. It is an ethical lapse because it values an imagined and uncertain harm to a given group above a commitment to scientific truth. It is a scientific lapse, because the probability of such harm is unknown and probably unknowable, at least over the long term. And, as the Lysenko story shows, much greater harm may come of suppression: abandonment of a broad commitment to free inquiry may have a substantial cost. Unfortunately, many accept an argument along Kitcher's lines as reason to discourage or even suppress research on the endogenous causes of racial and sexual disparities.[28]

There are other objections to the study of endogenous causes. One, of course, is its unfortunate history (see the Glasgow letter, quoted earlier). If there are real statistical differences in, say, cognitive ability between identifiable groups, then discrimination is justified. Well, no, it is not, and by now I suspect that most people have learned to behave better. This objection has now much less force than it once had.

Another objection is more subtle: "If there really are these ineradicable differences between people, so what? We can't do anything about them, so why study them?" Even a thoughtful critic like John McWhorter has made this point as his only negative criticism of Charles Murray's new book *Facing Reality*:[29] "Yet it's reasonable to ask of Murray: Why are you airing this information [about racial IQ disparities]? To what end?"[30] This point obviously appeals to many. Nevertheless, it is false, because if such differences really exist, the disparities they may produce will remain—and demand an explanation. If study of them is suppressed, the correct explanation will be unavailable. Human nature being what it is, if the real cause is ruled out, not by science, but by politics and groupthink, something else will take its place: welcome to systemic racism, invisible, unmeasurable, and ineradicable—and all the social division it brings in its train.

Racial disparities per se are uninterpretable. They should therefore play little or no role in public policy discussions. If they are used in political arguments, science has an overwhelming obligation to look for their causes, endogenous as well as exogenous.

PART 5

History of Science

CHAPTER 15

Neutral—Or Not?

"It is well-known that history of science is much hampered by the double obstacle that most scientists have no sense of history, while most historians are ignorant not only of the facts, but of the very spirit of science."
—*Robert S. Cohen and John J. Stachel*

Historians have always been subjects of criticism: charges of bias, omissions, inaccuracies, fabrications, insinuations, and just a bad attitude have been aimed at even the most eminent. Edward Gibbon, author of the magisterial *History of the Decline and Fall of the Roman Empire* (1782) was not immune. He was accused of disrespecting Christianity, for example in this passage:

> Our curiosity is naturally prompted to inquire by what means the Christian faith obtained so remarkable a victory over the established religions of the earth. To this inquiry, *an obvious but satisfactory answer may be returned; that it was owing to the convincing evidence of the doctrine itself,* and to the ruling providence of its great Author. But as truth and reason seldom find so favorable a reception in the world, and as the wisdom of Providence frequently condescends to use the passions of the human heart, and the general circumstances of mankind, as instruments to

execute its purpose; we may still be permitted, though with becoming submission, to ask, not indeed what were the first, but what were the secondary causes of the rapid growth of the Christian church [emphasis added].[1]

A modern reader will surely see Gibbon's acquiescence to the truth of Christianity as graceful, though possibly insincere. But he had a point about *proof*. Philosopher David Hume perhaps convinced Gibbon that the facts of science are provable but the "facts" of religion for the most part are not. They are matters of faith, not science. There is no *scientific* evidence for the truth of Christianity. So Gibbon was perfectly justified in looking for causes other than "convincing evidence of the doctrine" for the rise of Christianity and the fall of Rome.

History of science is different. Science's whole existence depends ultimately upon *convincing evidence of the doctrine itself.* We believe in the facts of science because they have passed our tests. These proofs surely are, in Gibbon's words, the *first causes* for the rapid growth of science. It follows that a prerequisite for any historian of science is *to understand the science.* Understanding science is essential for an accurate account of its history. As we will see, some contemporary historians of social science are not doing well in this respect.

History of science, any history, should also strive to be fair, unbiased: it should report all relevant facts, not just facts that fit a particular "narrative." But a bias need not spoil a history if it is frankly acknowledged. I begin with an example.

Science and Ideology—Overt

J. D. "Sage" Bernal (1901–1971) was a polymath Irish X-ray crystallographer and a committed communist. He was a prolific writer as well as an influential scientist who played a peripheral role in the Watson-Crick discovery of DNA as a senior colleague of Rosalind Franklin at Birkbeck College in London.[2] The Society for Social Studies of Science set up an award in his name in 1981.

In the early 1950s, Bernal wrote the first edition of the 1039-page, 4-volume book *Science in History*, which covers the history of science, very broadly defined, in a rather back-and-forth fashion (topics recur, widely separated) from the Stone Age to the twentieth century.[3] The book is generally well-regarded. Nevertheless, it was in the vanguard of a distressing trend.

Bernal certainly understood science and saw its value and its effect on society. But he saw also, perhaps too vividly, the reciprocal effect of society on science. In the preface he writes:

> In the last 30 years, largely owing to the impact of Marxist thought, the idea is grown that not only the means used by natural scientists in their researches but also the *very guiding ideas of their theoretical approach* are conditioned by the events and pressures of society [emphasis added].[4]

Bernal emphasized the science → society link in the early history, but tended to favor the reverse for more recent times. Proceeding from the axioms of Marx and Engels, which he thought of as certain truths, Bernal goes on to pronounce on various aspects of science and society. For example, "In the capitalist world the major feature of the 20th century has been the rapid growth to complete dominance of large combines, trusts or cartels, partly commercial, partly industrial."[5] Yes, unchecked capitalism produces monopolies. We see it now not so much in heavy industry as in Big Tech—Meta, Twitter, Amazon, and Alphabet—"natural" monopolies produced by huge positive feedback to scale, advantages not of production but in augmented demand: the most popular service will be most attractive to new users. And all facilitated by the eager purchase of potential competitors.

Of course communism doesn't give rise to monopolies; it doesn't need to because it is itself a monopoly. The state runs everything, which Bernal applauds. He seems unaware of Friedrich Hayek's influential 1944 *Road to Serfdom*,[6] written while Hayek was working at the London School of Economics, just a couple of miles from Birkbeck where Bernal was a professor. Hayek convincingly pointed out the difficult and perhaps insoluble information problems confronted by central state planning.

Marxism is intellectually imperialist: it covers every human activity. Bernal's belief in Marxism led him to consider Marxist social science a real science and to include in it everything from history to law. As for experimentation, which is the essence of science, the proto-communist societies of Soviet Russia, China, and Eastern Europe seemed to him just as good as the laboratory studies of Michael Faraday or Claude Bernard—an extraordinary delusion, a product of the devotion to Marxism and the Soviet "experiment," that led him to support the destructive pseudoscience of Trofim Lysenko, which I discussed earlier. Lysenko didn't believe in genes, favoring a sort of Lamarckism, the idea that acquired characteristics can be inherited, an idea that survives in modern biology only in etiolated form as in, for example, *genetic assimilation*.[7] Lysenko's failed, Stalin-supported, agricultural policy led to the deaths of millions. Dissenting scientists lost their jobs or were imprisoned or even in some cases executed by the Stalinist regime. Bernal remained faithful to Stalinist Marxism.[8]

Nevertheless. *Science in History* is a fascinating read, full of interesting facts and speculations. Bernal understands Darwin even as he deplores one kind of Darwinism:

> The simple tracing of evolutionary relationships between organisms and the building of elaborate family trees distracted naturalists from the study of the actual lives of the inner workings of animals and plants. . . . For this no one could blame Darwin himself, who was, as his detailed researches on such varied topics as earthworms, carnivorous plants, and the expression of the emotions show, one of the pioneers of experimental biology.[9]

Bernal blames eugenics not on Darwin but on the brilliant Francis Galton. "It was with the highest of intentions that Francis Galton, Darwin's cousin, set about studying the heredity of men of exceptional ability in Britain."[10] Scientists share some blame for the horrors that followed, says Bernal:

> [Scientists'] fear of entangling themselves in politics meant that they left the social application of their own ideas to other people, and made no effective protest against the perversion of the products of their own researches.[11]

Thus, James-Bond-like, scientists are handed a license, if not to kill, at least to politic. Others might see Lysenko as a textbook example of the dangers of politicized science, but his devotion to Soviet Marxism made this impossible for Bernal. Ideological commitment may make for lively writing; it is not a basis for truthful history—of science or anything else.

Science and Ideology—Covert: *Darwin and*
(Political) History of Science

Bernal well-understood science and his ideology was up-front. A more recent book, on Darwin, is less overt. It deals less in substantive criticism than in snide implication. And as Christian apologist William Paley said in response to Gibbon, "Who can refute a sneer?"

Sneers have been directed at Charles Darwin in influential modern histories. For example, there is some controversy about Darwin's reaction to Alfred Russel Wallace's scooping his discovery of evolution by natural selection. In 1858, Wallace, a young naturalist who had corresponded previously with Darwin, sent him a short paper. He asked Darwin to send the piece to Darwin's friend and mentor, geologist Charles Lyell. Wallace's paper reached the same conclusion Darwin had been buttressing with tireless research for the previous twenty years. Darwin was devastated. Some years earlier he had shown an essay describing natural selection to several colleagues, so there is no doubt that he had the idea first, not to mention a pile of data and arguments to support it and refute many possible objections.

Darwin had delayed publication for several reasons. Possibly he thought the anti-religious implications of evolution too inflammatory to publish without overwhelming support—Robert Bridges in his popular (but anonymous) *Vestiges of the Natural History of Creation* (1844) had achieved the kind of notoriety that was anathema to Darwin. More likely he was simply conscientious; he wanted to get it right before going public. In the end his friends Lyell and Joseph Hooker arranged for a joint paper to be presented that same year to the Linnaean society, although neither author was present—Darwin, sickly at home in Downe, Wallace still in the South Seas.

Wallace did not mention publication in his letter, but this is what Darwin wrote to Lyell: He would "of course, at once write & offer to send [Wallace's paper] to any journal." Yet "all my originality, what-ever it may amount to, will be smashed."[12] Still, Darwin added, "I should be *extremely* glad now to publish a sketch of my general views in about dozen pages or so. But cannot persuade myself that I can do so honourably."[13] These are not the words of a dishonest or self-serving man. In after years, Wallace was perfectly happy with the way Darwin had treated him and named his book on the topic *Darwinism*.

Yet, here is the sneer. We learn from Adrian Desmond and James Moore's 808-page *Darwin: The Life of a Tormented Evolution-ist* that "Irony and ambiguity shrouded Darwin as no other eminent Victorian" (an allusion, of course, to Bloomsbury groupie Lytton Strachey's brilliantly destructive caricatures of Victorian heroes such as Florence Nightingale and General Gordon[14]).[15] Desmond and Moore go on:

> [Darwin] hunted with the clergy and ran with radical hounds; he was a paternalist full of *noblesse oblige* [a bad thing, apparently], sensitive, mollycoddled [despite heroic actions during the *Beagle* voyage that got a channel named after him], cut off from wage-labour and competition [could he have done so much science otherwise?], who unleashed a bloody struggle for existence [blaming the messenger?]; a hard-core scientist addicted to quackery, who strapped "electric chains" to his stomach and settled for weeks at fashionable hydropathic spas [he tried rem-edies for a medical condition that had no cure at that time[16]] . . .[17]

Polite inquiries of a scientific opponent, the American Alexander Agassiz, are labeled "pickpocketing."[18] Describing Darwin's interac-tion with pigeon-fanciers, they write: "His kindly paternalism gave their backyard hobby a certain cachet. But to the end he remained impeturbably a gent among working fanciers."[19] Moreover, Darwin *dissimulated*. To his cousin Fox on the mutability of species Darwin wrote "I mean with my utmost power to give all arguments & facts on both sides."[20] Desmond and Moore know better: "Balance and doubt were a public mask. Despite appearances, he knew exactly what he was doing."[21] These eminent historians of science discern dishonorable motives even in Darwin's most transparently decent actions. It seems to be his upper-middle-class origins they particularly dislike, irrelevant though these should be to his scientific achievements.

Wallace and Darwin viewed natural selection a bit differently. According to Desmond and Moore, Darwin provided what his supporter T. H. Huxley wanted: "a new competitive, *capitalist* sanction in place of Anglican Oxbridge paternalism [emphasis added]."[22] A "Malthusian, capitalist, competitive mechanism . . . unlike any rival evolutionary theory."[23] Darwinian evolution was a metaphor for industrial capitalism. "Nature was a self-improving 'workshop,' evolution the dynamic economy of life. The creation of wealth and the production of species obeyed similar laws."[24] Darwin's views are hardly surprising, they suggest, as "[He] was a heavy investor in industry," not to mention family connections to the porcelain-rich Wedgwoods.[25] Never mind; true or false, Darwinian natural selection is just capitalism by another name.

Wallace, on the other hand, was self-employed, more interested in cooperation than competition and a "self-taught socialist." "Wallace's naturally selected group morality was leading society in a very un-Darwinian direction [not that Darwin ever speculated about society]."[26] Wallace also believed that humanity, because of language and intellect, was special. Wallace believed in the inevitability of evolutionary progress. Darwin did not; contemporary evolutionary biology also does not.

Needless to say, Desmond and Moore do not think that Darwin treated Wallace well. He should have seen from their earliest correspondence that Wallace was on to natural selection: "He did not really catch Wallace's drift."[27] Darwin should just have bowed out, apparently. This gross misapprehension has made it into popular writing, even of usually accurate and always-entertaining Tom Wolfe.[28]

The authors seem to trace Darwin's ideas not to his own genius and the mountains of data he gathered, but to his class, his upbringing, and his position in society. How much more likely is it that their own biases arise from a like cause! Quite possibly the authors' sympathy for Wallace follows from their own political preferences. Previous coverage of Darwin "served a purpose a century ago in securing Darwin's immortality. But today's needs are different. . . . We want to understand how his theories and strategies were embedded in a reforming Whig society."[29] *Today's needs?*

Desmond and Moore are determined to tie Darwin's behavior to social issues rather than to his curiosity and love of nature, puzzling over rocks and atolls, barnacles and beetles. In a subsequent book, *Darwin's Sacred Cause*,[30] they link his theory to his abhorrence of slavery—an intriguing connection since at least one other, admittedly odd, book makes Darwin a racist.[31] The man's intellectual independence, his ability to ponder problems invisible to many others, is lost in Desmond and Moore's account. Society, social movements, social agitations, *social* is the thing. Yet, all admit that Darwin was for most of his life a recluse. He communicated with others largely by letter and largely about biology. He wrote almost nothing about society. He was neither a sociologist nor a political scientist; he was above all a biologist. His inspiration was from nature, not society. This is a conclusion that Desmond and Moore seem unwilling even to entertain, much less accept.

Desmond and Moore's book is long and copiously referenced. They (unlike some others I will mention) obviously understand the science. But they give little weight to it. Instead their view of the man and his achievements is suffused with a sneering, politically tinged tone that disparages Darwin and devalues his scientific contribution. Darwin was just a predictable product of his time and place. No room for talent or even luck. No room for the verifiable truth of the man's ideas. The book is not a history of science so much as science as politics by other means.[32]

CHAPTER 16

Historians of Science or Political Journalists?

N aomi Oreskes is a distinguished Harvard historian of science. Trained as a geologist, her specialty is climate science; she has long been an advocate for AGW.[1] Testifying to her own open-mindedness, she has adopted what she calls the Red State Pledge: "If I get invited to a Red State, I do everything in my power to accept that invitation," feeling that she might learn something from the people of, say, South Dakota that she might not get at, say, Princeton. But, like Professor Bernal, she doesn't think that science should be value-free. Speaking of Mary Shelley's *Frankenstein*, she says that without values, "The science is the monster."[2]

But of course, "values" can be a problem if they interfere with the evaluation of evidence. A passion for truth is one thing: a passion for social justice quite another. The major general-science journals—*Science*, *Nature*, and *Scientific American*—have all become proudly political in recent years ("Science and politics are inseparable"[3]), and the trend is evident in Oreskes and Erik Conway's influential 2010 book *Merchants of Doubt*.[4] The book is a full-throated attack on critics of the received scientific wisdom on a number of environmental issues: acid rain, the ozone hole, global warming, and—the topic on which I will expand—secondhand (environmental) tobacco smoke (ETS).

The book is a "riveting piece of investigative reporting" says a *Guardian* reviewer, who goes on to say, "The far right in America, in its quest to ensure the perpetuation of the free market, is now hell-bent on destroying the cause of environmentalism," which signals the book's genre, its political angle, and its intended audience.[5]

Merchants of Doubt is journalism, to be judged on how attention-grabbing it is, with accuracy and comprehensiveness as subordinate virtues. Like much journalism it embraces the genetic fallacy not as a fallacy but as a useful dramatic tool: Never mind what he says, why does he say it? Who's paying? It assumes, as many journalists do, that peer review is a guarantee of truth, rather than a weak barrier against obvious mistakes—and sometimes a suppressor of dissent.

The book's coverage is partial: the most prominent AGW "deniers" are elderly senior scientists. Not included are younger, more active scientists less likely to be dismissed as "past it."[6]

A closer look at the case of secondhand smoking, and what science actually says about it, will help to dispel the book's mythical view of science. Science is less univocal than activists want it to be. That is no less true of secondhand smoke than of climate change and other controversial scientific issues.

Secondhand Smoke

Any issue related to health risks generates a controversial and tendentious literature of research, "research," and opinion. Secondhand smoke, with its links to a large industry whose motives will always be suspect, is no exception.

Detecting a meaningful health effect of environmental tobacco smoke is bound to be very difficult. Any effect on mortality is likely to be modest (only a small fraction of the exposed population will be affected), and the effect will be delayed. If people are going to get sick or die after exposure to ETS it will only be after a lapse of time, and not hours or days but years. Proving an effect of ETS is tough.

Experiment, the only way to establish causation beyond a reasonable doubt, is impossible, for both ethical and practical reasons: we cannot expose people to things that may make them sick, and even if we could, there will be few researchers eager to take on a project with an outcome necessarily delayed for decades. Measuring the supposed ill effects of ETS is a trans-science problem.

Nevertheless, Oreskes and Conway feel that a harmful effect of a low concentration of tobacco smoke is "common sense": "The U.S. Department of Health and Human Services tells us that 'there is no risk-free level of exposure to second-hand smoke: even small amounts . . . can be harmful to people's health'" (quoted from the 2006 surgeon general's report). They, and probably a majority of medical people, have no doubt "that secondhand smoke can kill."[7]

But is it true in any meaningful sense? After all, everyone dies, with no intervention at all, and the effects of ETS, if any, are delayed, so what we are looking for is not immediate lethality, but some shortening of life.

Since experiments are ruled out, we are left with epidemiology as our only guide. Epidemiology has many limitations, the most obvious being that it can only detect correlations, which may be causal but may not. As I point out in the appendix, there is a good correlation between the number of divorces each year and the number of apples imported to the United Kingdom, and a strong correlation between spider bites and word length in a national spelling bee.

Despite many absurdities like this, "correlation is not causation" is a truism as often ignored as acknowledged.[8]

Ad Hominem?

Oreskes and Conway accept epidemiological data without demur. They begin their discussion with the surgeon general's confident conclusion: secondhand smoke can kill. But it is soon clear that their target is not so much tobacco as the tobacco companies:

> Just as the tobacco industry knew that smoking could cause cancer long before the rest of us did, they knew that secondhand smoke could cause cancer, too. In fact, they knew it well before most independent scientists did.[9]

These statements refer to a 2006 RICO judgment which concluded that the tobacco companies were guilty of racketeering for claiming that ETS and low-tar-and-nicotine cigarettes have smaller toxic effects on human health than mainstream smoke from regular cigarettes.[10] As for the tobacco companies knowing about the dangers of cigarettes in advance of the public, that is surely nonsense: The definitive publication was the research by British epidemiologists Richard Doll and Bradford Hill in 1950,[11] and well before that time cigarettes were called "coffin nails" and soon after "cancer sticks."[12] People were well aware of the dangers of smoking.[13]

The politicization of the science related to ETS is quite extraordinary. As we will see, the anti-smoking side has simply ignored substantial data on the other side.

I begin with the *Merchants of Doubt* account. The weight of their claim that ETS is lethal rests on a 1981 Japanese study by Tekeshi Hirayama.[14] A 2011 summary of his work by D. R. Smith and E. J. Beh cites a 1979 paper by James E. Enstrom (remember that name), which includes a comment on "a large relative increase in lung cancer cases in non-smokers in the United States that was observed between 1914 and 1965."[15] Enstrom's observation apparently alerted Hirayama to the problem. Hirayama's study looked at the wives of smokers and nonsmokers in Japan, hypothesizing that the wives of nonsmokers, free of chronic exposure to ETS, would have lower lung-cancer rates than the wives of smokers, long exposed to ETS.

Ideally, one should focus on mortality rates, because these are more reliable than, say, diagnoses of emphysema or cancer. Unfortunately, Hirayama looked not at death per se, but death from lung cancer. He studied a large population: "These cases occurred among 91,540 non-smoking married women whose husbands' smoking habits were studied," a subpopulation selected from a total of 265,118 adults.[16] (For some reason *Merchants of Death* refers to "540 women," suggesting that Oreskes and Conway misread the study.[17])

The aggregate figure for lung-cancer deaths among the wives of smokers appears in the methods section. Apparently 245 married women in the sample died from lung cancer, 174 married to non-smokers died, so 71 nonsmoking wives married to smokers died from lung cancer. Table 1 in the results section shows more details. Sorted by husband's age and the amount he smoked—whether 0 cigarettes per day, 1–19 cigarettes per day, or more than 20 cigarettes per day—there is a clear trend: the more the husband smoked, the greater the wife's cancer risk.[18]

Statistical objections were raised to this study,[19] and the Smith and Beh summary paper admits that the aggregate data do not pass the chi-square test of significance. But, by parceling out the data by age and considering three variables (husband's age, cancer status, and smoking frequency) rather than just two, these authors are able to get a "highly significant" result.[20] (Why the husband's age rather than the wife's is critical is not explained.) Smith and Beh conclude that Hirayama was, after all, correct in claiming an increased cancer-death risk from secondhand smoke. As to the detailed assumptions underlying this now more complicated statistical model, little is said. Yet, as I point out in the appendix, these assumptions are as important to the conclusions as the actual data, even in studies much simpler than Hirayama's. Nevertheless. Hirayama's results were, and are, generally accepted.

There is a 2003 study, not mentioned either by *Merchants of Doubt* or the summary authors, that might have settled the issue—not by proving that ETS is harmless (which is impossible), but by showing that a well-designed, large-scale attempt to find harm failed to do so. The study is by the aforementioned James E. Enstrom and coauthor Geoffrey C. Kabat, again in the *British Medical Journal*.[21] (This reference appears in the citations for chapter 7 of the 2006 surgeon general's 709-page report but, curiously, is not discussed in the text;[22] it has vanished from the 727-page 2010 report.[23]) The authors looked at 35,000 never-smokers in California with spouses with known smoking habits. The participants were selected from 118,000 adults enrolled in late 1959 in a cancer-prevention study. The numbers are large, and the researchers asked a simple question: Are Californians who are married to smokers likely to die sooner than those married to nonsmokers? Their answer is unequivocal: "The results do not support a causal relation between environmental tobacco smoke and tobacco related mortality, although they do not rule out a small effect."[24] Enstrom and Kabat found no, or little, effect of secondhand smoke. The study received some criticism, but more about its sources of funding than its scientific details.[25]

In sum, we don't know whether secondhand smoke is harmless. What we do know is that a persuasive study that failed to find harm has been ignored by Oreskes and Conway and airbrushed out of the surgeon general's official report.

There are strong feelings on both sides of any smoking-related issue, although the anti-smokers seem to have the edge. Most smokers acquiesce because they are scared and wish they could quit. Nevertheless it is hard to conclude from Enstrom and Kabat and other studies[26]—like the "natural experiments" provided by the before-and-after analyses of cancer rates following smoking bans[27]—that the effect of ETS is anything other than trivial. It is also hard not to conclude that there is considerable bias against anything that seems to minimize the dangers posed by smoking or concedes anything positive about the tobacco industry. Figures on the wrong side of the anti-smoking argument such as respected science writer Gina Kolata and scientists Fred Singer and Fred Seitz are routinely demonized. Research sponsored by the tobacco industry (for example, Alan Rodgman and Thomas A. Perfetti's book *The Chemical Components of Tobacco and Tobacco Smoke* and related papers[28]) is ignored.

It is tough to give a balanced picture of the smoking landscape. But Oreskes and Conway don't even try.

No Threshold?

There is one more arrow in the anti-tobacco advocates' quiver: something called the *linear no-threshold* (LNT) assumption.[29] The argument goes like this: There is little doubt that heavy cigarette smokers are at risk of lung cancer and related diseases like chronic obstructive pulmonary disease (COPD). We "know" that the effect of a toxin is strictly proportional to its concentration, with no lower bound, no threshold. Hence the lethality of even low doses of tobacco smoke is proved: "There is no risk-free level of exposure," as the surgeon general said. Which opens the door wide, justifying restrictive regulation and confiscatory taxation on smoking and smokers. Pubs in England, and the social space they provide, have been decimated by bans on all smoking in interior public areas, an unmeasured social downside.[30] Many UK citizens now spend their time smoking and watching a screen at home, rather than discussing the issues of the day with their friends over a pint. This may please social media bosses; it is hardly a benefit to human happiness—or to democracy. The human costs of regulatory impositions do not figure in the calculations of Oreskes and Conway and other anti-smoking advocates.[31]

Is the LNT assumption correct? Is there really no lower bound to the damage caused by toxins like ETS? Could LNT be true of all toxins, as its adherents claim? Or is it possible that each toxin has its own *dose-response curve*?[32] The no-threshold assumption does after all go against the (admittedly self-serving) saw "the solution to pollution is dilution" and the old adage quoted in *Merchants of Doubt* "the dose makes the poison."[33]

Well, yes, this *does* seem like common sense. Many substances, salt or alcohol, say, are toxic in high doses but harmless or even beneficial in low ones. Is ETS different? Even if true for some toxins, LNT surely cannot be true of all. And the human body, all bodies, are what Nassim Taleb has called *antifragile*, they respond to stress by becoming stronger.[34] The immune system, after all, requires some exposure to infectious agents to develop a resistance to them. Too much may be lethal, but no exposure at all can also have harmful effects.[35] There is no a priori reason to expect that the effects of all toxins are strictly proportional to dose, or even monotonic.

An EPA paper points out that if we understood the "mode of action" of a potential pollutant, the way in which it affects the body's biochemistry, there would be no doubt about its effect at any dose.[36] Unfortunately, even today, we rarely do. The way that ETS interacts with the body's physiology and biochemistry is still largely unknown.

There was much criticism of the no-threshold idea. The LNT assumption is false as a general truth. Research on radiation risk, where the idea originated, has come under increasing attack. Bearing in mind the difficulty of proving a negative, clear no-harm radiation thresholds have been found in animal and plant preparations. The chequered history of honest fact in private letters and dishonest denial proclaimed in public is recounted in detail by Edward Calabrese, who concludes that the no-threshold hypothesis is invalid for radiation risk.[37]

The bad effects of radiation are relatively rapid, and the use of nonhuman subjects allows for experiment.[38] Comparable research on ETS with human subjects is prohibited by ethical and practical constraints. In the absence of conclusive human data, the convenient LNT assumption has been applied by fiat to all potential carcinogens, including tobacco smoke, where it is almost certainly wrong.

Precautionary Principle

The LNT assumption has a sad history in which its inherent implausibility and weak empirical basis have given way before the regulatory imperatives of the *precautionary principle*,[39] which goes well beyond cost-benefit analysis and puts the burden of proof on the manufacturer: "guilty until proven innocent," in other words. Despite the problems and criticism, the no-threshold assumption has become an EPA rule, at least as far as potential carcinogens are concerned. With the LNT as a superweapon, there are few limits to the controls that may be imposed in the interests of "health and safety." Petrified by proclamations that there is "No safe level of smoking: even low-intensity smokers are at increased risk of earlier death,"[40] the populace wilts before regulations based not on a solid foundation, but on a beach of sinking scientific sand washed by the waves of politics.

What to Do When the Science Is Uncertain?

Controversies in this area typically arise at the margin. No one these days argues about the cancer risk to which heavy cigarette smokers are exposed. But when the science is ambiguous, when the health effects of a toxin—especially one like ETS that is experienced at low levels—when its effects have not, and perhaps cannot, be established with certainty, then the issue changes. It becomes a matter to be decided by not by science, but by ethics and the political process. In partial recognition of this shift, the EPA in 1991 decided to treat cancer and non-cancer risks differently:

> In the absence of clear evidence to the contrary, EPA assumes that there are no exposures that have "zero risk"— even a very low exposure to a cancer-causing pollutant can increase the risk of cancer. . . . EPA also assumes that the relationship between dose and response is a straight line—for each unit of increase in exposure (dose), there is an increase in cancer response.

For non-cancer risk, on the other hand:

> A dose may exist below the minimum health effect level for which no adverse effects occur. EPA typically assumes that at low doses the body's natural protective mechanisms repair any damage caused by the pollutant, so there is no ill effect at low doses. . . . The dose-response

relationship . . . varies with pollutant, individual sensitivity, and type of health effect.[41]

I have been unable to find a scientific justification for this distinction between cancer and non-cancer risk. In 2005 the EPA published
Guidelines for Carcinogen Risk Assessment, which should say something about the issue, but says only, "The Agency's more current
guidelines for these effects [that is, thresholds] . . . however, do not
use this assumption."[42] The 166-page document reads like a "how-to"
manual on scientific research. It attempts to anticipate every problem
and confound, for a wide range of toxins, tumor types and subject
populations—an impossible task. Science doesn't work like that. In
practice each problem must be tackled with an open-ended ingenuity that defies detailed advance specification. The attempt to specify
every research direction leads, as this document does, to a tree with
endlessly proliferating branches—to confusion. This is not how science should work.

This approach almost guarantees omissions and even errors. The
document says little about the problems caused by delayed effects,
for example, and seems hesitant even about the meaning of causation:
"A causal interpretation is strengthened when exposure is known to
precede development of the disease."[43] Just "strengthened"? Since
when is a cause following its effect even a possibility? These cancer
guidelines say nothing about why the threshold issue is different for
cancer and non-cancer risk. The distinction seems to derive simply
from the special scariness of cancer.

The EPA itself seems to have set itself an impossible task: to
understand the science of toxicity while coming up with rules and
regulations to control it. Perhaps the scientific mission of the EPA,
indeed of all health agencies, should be severed from its regulatory
function: actions separated from science, passion from fact. Proposed

new guidelines may tighten up the factual basis for regulation, but the way the regulations are written and vetted should also be looked at afresh.

When experiment is impossible, scientists, health-policy officials, and science journalists should above all be cautious in their actions and proclamations. Very often they are not.

As I have pointed out here, Oreskes's case against passive smoking is flawed: a critical study is missed, some studies are misrepresented, and the book seems oblivious to the scientific difficulties involved in studying potentially dangerous effects that are likely to affect only a small fraction of the population and only after long delay. Nevertheless, it is in many ways a brilliant and persuasive book.

The case might end there, but it does not, because Naomi Oreskes is not a journalist but an academic historian of science. Has scholarly history of science now become nothing more than political journalism—with footnotes? Probably, as Oreskes, discussing her most recent book, happily promotes shared values as a way to get a scientific message across: "Scientists need to talk about the values that motivate them and shape the science they do."[44] Do they, really? Darwin would not have agreed.

First, Understand

Nancy MacLean is the William H. Chafe Professor of History and Public Policy at Duke University. She has written a book whose central figure is a soft-spoken Southern economist, winner of the 1986 Nobel prize in Economics, James McGill Buchanan.[45] One critic of the book, Michael Munger, a Duke colleague and expert in *public choice*, Buchanan's field, calls it a "remarkable book,"[46] albeit "speculative historical fiction." A strong charge, but amply documented in Munger's long review.[47]

Democracy in Chains is about politics, about threats to American democracy. It is written in a lively, whodunit fashion as a sort of spy story. James Buchanan is the evil conspirator at the heart of the plot. Buchanan died in 2013, not long before MacLean took up her work. Conveniently, as the timing meant that she could freely attribute motives to him without danger of rebuttal; and she found a trove of Buchanan's papers, then un-curated and uncensored, in a house at George Mason University.

From this "unlisted archive," she has contrived a sinister plot. Buchanan, it turns out, aided and abetted by "shadowy billionaires," most notably the Koch brothers (MacLean seems to assume that her readers will share a certain view of *them*) wanted to destroy democracy. She can now reveal "the utterly chilling story of the ideological origins of the single most powerful and least understood threat to democracy today: the attempt by the billionaire-backed radical right to undo democratic governance":

> Buchanan . . . argued that representative government had shown that it would destroy capitalism by fleecing the propertied class—unless constitutional reform ensured economic liberty, no matter what most voters wanted.[48]

This single sentence captures MacLean's innocence of political economy. She doesn't understand the science.

"Representative government"—democracy—means rule by the people. But just what people? It has been known since at least the eighteenth century that if the franchise is broad, including many poor people as well as a few rich ones, then it has a potentially fatal flaw:

> A democracy is always temporary in nature. . . . A democracy will continue to exist up until the time that voters

discover that they can vote themselves generous gifts from the public treasury. From that moment on, the majority always votes for the candidates who promise the most benefits from the public treasury, with the result that every democracy will finally collapse due to loose fiscal policy, which is always followed by a dictatorship.

The source of this quote is uncertain. (Eighteenth century Scottish legal figure Alexander Tytler is the most popular source, but many others have made similar observations.) But its truth is undeniable. A broad democracy without checks and balances to protect wealthier minorities from expropriation by the state is an unstable arrangement. Historian MacLean seems to have missed *Federalist* number 10.[49]

American and British democracies two or three hundred years ago were not, of course, "broad." For much of the time, only those with skin in the game—men, owners of land, or payers of tax—were eligible to vote. (The system was then subject not to the Tytler dilemma, but its opposite: exploitation of the poor by the rich.) Now, however, the United States has a very broad democracy: there are no sex or property qualifications, the voting age (eighteen) is below the legal drinking age (twenty-one)—and a few politicians want to lower it even further. Without checks and balances to limit total dominance by the majority, the system would indeed be unstable.

There are of course many such checks in modern America, in the Constitution and after. MacLean describes a few and deplores all of them. But that some such restraints are necessary is unarguable. Even thoughtful capitalists deplore monopolies. In MacLean's majoritarian version of democracy, the majority *is* a monopoly, with absolute power. She is surely aware of Lord Acton's dictum about absolute power. In fact, as Munger points out, she doesn't even recognize this elementary problem. Public policy professor MacLean seems not to understand one of the most basic facts about democracy.

It is always possible, given a large enough collection of facts, to select from them just those that support a desired story. The most infamous example is Howard Zinn's *People's History*,[50] the most recent, Nikole Hannah-Jones's *1619 Project*.[51] It is also what Desmond and Moore did in their *Darwin*. They ignored the data, the fossils, the specimens, and Darwin's many curious speculations about what he was finding. What was left was the social milieu in which Darwin found himself. *Aha! That* must be what drove him to his atheistical, capitalist solution they conclude. Well, no: what they left out is much more important than what they retained.

Nancy MacLean's book is in this same tradition. Buchanan in his many writings and talks always emphasized at the outset that *ideas* were supremely important to him. His overriding motivation was to understand/explain (his usage) how individuals make choices and how different social arrangements affect their choices: "We shall construct, in an admittedly preliminary and perhaps naïve fashion, a theory of collective choice."[52] Yes, he was a libertarian, or a *classical liberal*, as he preferred. Yes, he would no doubt have liked society to be organized differently. But no, these desires were not the main source of his work. He admitted in one interview that a socialist society would be perfectly acceptable to him if it were the result of genuine consensus. Yet Buchanan's supposed political preferences absolutely dominate MacLean's account. Buchanan is transformed from a thoughtful intellectual to a sinister hatcher of plots.

Michael Munger's detailed critique points out the many places where MacLean has found conspiracy when Buchanan was simply advancing an intellectual movement. For comparison, he quotes MacLean's Marxist colleague Fredric Jameson, who has for many years professed his intention to create a "Marxist intelligentsia."[53]

Buchanan's half-serious aim was to create "an effective counterintel-ligentsia."[54] The briefest of glances at the political preferences of social-science faculty in American universities shows that Jameson has come a lot closer to achieving his goal than Buchanan. So much for the "far-flung . . . intricately connected institutions funded by the Koch brothers and their now large network of fellow wealthy donors."[55]

In his Nobel lecture, Buchanan pointed out the eighteenth-century discovery of Adam Smith's "invisible hand," the way that individual, self-interested choices can lead to collective good.[56] Public Choice Theory, a product of work by Buchanan and his long-time collabora-tor Gordon Tullock,[57] attends to "the processes through which indi-vidual choices are exercised."[58] Basic economic principles can in this way be extended to a variety of institutional settings. This ultimately voluntary basis for political agreement runs counter to the postmod-ern/Marxist emphasis on politics as power. Instead of being passive objects of power, public choice sees individuals in a democracy as voluntarily obeying rules arrived at through an agreed process. It is an attempt to extend Smith's insight to collective action. Since claims about unfair power relations are at the heart of progressive politics, MacLean is perhaps unwilling to consider an approach that tries to understand democracy as the outcome of voluntary action by free individuals.

MacLean's book has received considerable acclaim. The first few pages are devoted to no less than twenty-seven enthusiastic endorse-ments, ranging from Bloomberg to the *New York Times Book Review* and *The Nation*, from the UK *Independent* to Oprah. There is an enthusiastic review in *The Atlantic* (a "vibrant intellectual history" the book is "part of a new wave of historiography").[59] Why this enthu-siasm from so many left-leaning sources? The answer is simple. First, the book is readable, well written, and has a great plot. But more importantly, it ticks every box of the progressive checklist, taking approved positions on all hot issues: capitalism (never defined), white men, and selected billionaires; the "radical" right, vouchers, and public schools; Augusto Pinochet, Scott Walker, the Mont Pelerin Society; anthropogenic global warming, Southern racism, and racial segregation. "Dog whistles" and "coded meanings" abound: individual liberty = capitalism, special interests = black people, criticism of "big government" = racial anxiety. The book is full of translations of innocent words into incriminating ones. If you are of the appropriate political persuasion, what's not to like?

In fact, the book is novella of the "based-on-real-events" type, masquerading as disinterested research. It is a textbook example of the dangers of science historians who "talk about the values that motivate them and shape the science they do..." Their work very soon descends from real history to propaganda with footnotes.[60]

Epilogue

Despite almost two years of public confusion, it was not my intention to end this book by saying anything at all about the still-ongoing COVID-19 pandemic. It still seems to be too early to draw conclusions; the supposed facts are sometimes contradictory, and passions still run high. This book is about problems with science, and the most important science in relation to COVID-19, molecular biomedicine (specifically mRNA and DNA technology), has in fact done well. Several adequate vaccines have been developed, surprisingly quickly, and at the time of writing the pandemic seems to be, if not waning, at least becoming less lethal.

On the other hand, COVID-19 epidemiology has gotten mixed reviews. Epidemiology does many things, but during an epidemic its main duty is to predict the future course of infection, to guide mitigation in the form of hygiene, masks, and different degrees of social isolation. Epidemiology predicts in two ways: one is by comparison with earlier outbreaks of a similar infection. Here the most comparable epidemic in the United States is the 1918–19 influenza pandemic, which lasted from eighteen months to two years with three peaks.[1]

(The bubonic plague of 1665, a more lethal disease with a very differ-ent mode of transmission and a different population distribution, nevertheless lasted a similar amount of time.) Mitigation measures in 1918–19 were similar to today, though less centralized.[2] There are vaccines for COVID-19, but there were none for the 1918 flu; so we should recover faster from COVID-19? On the other hand, the U.S. population now is both more mobile and more risk-averse than the population in 1918 (air bags, seat belts, and infant seats in cars; crash helmets on bicycles, motorcycles, and even toddler tricycles; no free-range kids, et cetera—none the case in 1918) and is therefore more reluctant to abandon mitigation measures than it was in earlier days. Conclusion: on balance, the COVID-19 pandemic will last lon-ger than the the pandemic of 1918.

Epidemiology has another method of prediction: computer-based models of the spread of infection.[3] Epidemiological models share the problem of climate models: they cannot easily be tested. Because every epidemic is different, in effect the prediction *is* the test. For that reason alone, we may expect them to be unreliable. A model from Imperial College, London, achieved fame, or notoriety, early on. It projected a worst-case scenario (2.2 million U.S. deaths) alarming enough to cause lockdowns that seemed to many both unnecessary and economically damaging.[4] Nevertheless, the model claimed to be partially vindicated when "infections went down steadily during lockdown, as predicted, and at the time of writing [October 2020] are rising again, just as pre-dicted."[5] Other models (a total of *thirty-nine*) may have done better.[6] The range of predictions is wide, however. The most recent bunch predicts "that the number of newly reported COVID-19 deaths will likely decrease over the next 4 weeks, with 4,400 to 12,900 new deaths likely reported in the week ending March 27, 2021."[7] A three-to-one range of outcomes is not the kind of accuracy we expect from other branches of science. What is the problem?

The problem is the quantity and accuracy of the information necessary for any model to make an accurate prediction. A brief look at how epidemiological models work illustrates how difficult it is to fill in the numbers the model must use.[8]

First, imagine the human population to be like randomly inter-acting gas molecules in a chamber. Begin with a population of H (healthy individuals). Then introduce one infected person (I). This model—any model—must then show how the numbers of healthy individuals, infected individuals, recovered individuals, and fatalities changes over time.

A simple assumption is that the individuals (molecules) move at random contacting each other with probability p in each time step. With another probability, q, contact results in passing an infection from I to an H: $I(t+1) = I(t)+1$, $H(t+1) = H(t)-1$. The product, pq, is the probability an infection is transmitted in each time step. Each infected individual either recovers, reverting to a healthy but immune individual, M, or dies, D. (From this can be calculated the average number of people infected by one infected individual, usually called R_o, a label which gives it a spurious stability—probability q must change as the number of infected and immune individuals increases, for example.) This process continues until the proportion of infected individuals stabilizes, hopefully at a low level. An obvious implication of any model like this is that reducing the value of p—probability of an encounter—must slow the process and thus predict that infections will go "down steadily during lockdown" and vice versa. Since almost any model must make this prediction, it is little help in choosing among them.

Notice the huge oversimplifications of this picture. Individuals are assumed to be identical: all have the same probability of making a contact—just like gas molecules. A realistic model would need either to abandon this assumption or break down the real population into units and categories small enough for the assumption to be valid. The time for an infection either to cure or kill is assumed to be the same for all, which is obviously not true. Perhaps an average could work, but how to compute it? And the time-to-outcome—kill or cure—begins with the moment of infection, which will not be the same for all. So the simulation must somehow either keep track of everyone, or make another uncertain averaging assumption.

Given the informational uncertainties, existing models seem in fact to have done about as well as could be expected. But answers to questions like *will "slowing the spread" only delay later disaster?* or *will delay lead to more or less mass ("herd") immunity?* remain uncertain. To the question of the proper balance between the benefits of lockdown and its economic, psychological, and educational costs, science has no ready answer. These questions remain a matter for ethical and political judgement. Cries to "listen to the science" may signal virtue. But in this case, there is no "the science," but only a range of findings, some more certain than others, often in conflict, and both always subject to revision. Science is rarely certain, especially in the midst of a novel epidemic. We should be very suspicious when it is invoked as an incontestable canon of divine wisdom.

Two other large topics had to be omitted from this volume: one because it is technical and would take much space to explicate, the other because it is more a political than a scientific issue. Economics is in a bit of a crisis. The subject began with Adam Smith in the eighteenth century as a combination of money- and wealth-based data combined with a mixture of human motives, from greed to empathy ("sympathy" in Smith's terms). As the mathematical method began to take over in the nineteenth century, the quantifiable money variable began to dominate, and other kinds of human motivation were sidelined. Along with math came a predilection for physics-sounding concepts like "multiplier," "elasticity," "friction," and "efficiency" that applied to more or less intangible social concepts. "Comparative statics," based on the assumption that underlying processes tended toward equilibrium, became a major area—even though both history and a little thought raise doubts that capitalist economies are ever truly in equilibrium, that is, static. The real, practical problem is understanding the dynamics that underlie booms and crashes; assuming "statics" begs the question. Supply-demand theory was treated as causal rather than functional (as in optimality theory) and often failed to predict accurately, which prompted a half-hearted correction in the form of behavioral economics. To tease out all these elements would require another volume.[9]

The other issue I have had to omit is that of sex and gender. The biological facts are clear. All mammals reproduce sexually; reproduction requires an egg and a sperm, the male supplies the sperm and the female the egg—no room for a third party. Male and female is it. A few nonmammalian species show surprising flexibility. Some zooplankton are parthenogenetic: "No sex, please, we're rotifers!" Adult swordtails (*Xiphophorus helleri*) a popular live-bearing aquarium fish, sometimes change sex from female to male. Many other fish species change sex as adults—from female to male and the reverse. The change is triggered by, for example, the composition of the fish school (if all female, one female may change to a male) or simply age.

Mammals are less flexible, but essentially all species that have been carefully studied show departures from normal heterosexual behavior—"normal" in the sense that it is both the majority pattern and essential to survival of the species. Human beings obviously also show variety of abnormal patterns such as homosexuality (As *Myra Breckinridge* author Gore Vidal once put it, "Homosexuality is natural. Notice that I did not say 'normal.'"[10]) and the wish to be, or feeling of being, the opposite sex (transgender). These variants form a small fraction of the population. Between 2 and 10 percent of the male U.S. population may be homosexual, and perhaps 1 to 2 percent of females.[11] The proportion of bi- and trans-sexuals is probably much lower. All these figures are moving targets and vary with the winds of cultural change.

The causes of the different variants are uncertain; doubtless both cultural and biological (genetic, developmental) factors are involved in a still-unknown mix. The adaptive function of nonreproductive variants is also unclear. Obviously if a substantial fraction of the population abandons their reproductive role, the future for the species looks dim. On the other hand, division of interests and the resulting division of labor may mean that some nonreproductive minorities contribute to rather than diminish the "fitness" of a culture. (Celibate religious sects are an obvious example.)

How variants should be valued and labeled is a matter for ethics and politics, not science. But biology matters and here politics can warp science. That there are two sexes is biologically indisputable. That there are several minority variants is also indisputable; but trans women are not actual females and sex is not a continuum.

There is a majority population of normal males and females: How should they be treated? There are differences between men and women. Some claims achieve a consensus, and almost everybody will agree to them; others not so much. For example, men are in general stronger than women: they run faster, can lift heavier weights, are better football players. They are more likely to commit crimes, especially violent crimes. Despite the miraculous Simone Biles, men can even be better gymnasts. Most people agree on these things. What do they mean for women's participation in the military? Trans-women's participation in female sports?

Psychologists also tell us that men tend to be more interested in things, women more in people. (There are of course many individual exceptions.) Can this difference partly explain male-female disparities in professions such as nursing and engineering? Males in most dimensions are more variable than females. On some cognitive measures, for example, the population means are the same, but the male distribution is more spread out (higher variance): there are more male idiots and geniuses. Alluding to this as an explanation for the disproportionate representation of men in math-related departments at Harvard got then-president Larry Summers into trouble a few years ago.[12] It is a fact, nevertheless.

Passionate advocates for "gender equity" are likely to question if not ignore biological male-female differences. Sex differences very quickly become political, not scientific issues. Equality before the law is a consensus view; but whether people should be equally treated in all other respects is a matter where, at least in recent years, proven and relevant biological differences have been sidelined. Sociocultural (gender) differences between males and females represent a bargain struck between gender-nonconforming individuals and society. The "norm" therefore changes as the culture changes. Although some gender questions may be scientifically undecidable, political, social, and ethical decisions that ignore biology are unlikely to be wise.

Postscript

Professor Jerry Coyne in a blog on the topic of secular humanism and religion implied that personal values about things like gay marriage, Christian symbols on public property, and secular blasphemy are behind my argument against his separation of secular and religious morality.[13] I do feel that no one should be penalized, much less lose his job,[14] for using a word, any word, with no bad intent. Coyne seems to agree. Beyond that I commit in this book only to those beliefs and practices that make science possible: curiosity, honesty, reason, open data and open debate, the ability to separate fact from passion, and the faith that these things allow us to discover truths about an orderly natural world.

Appendix:
The Replication Crisis and Its Offspring

Nonstatistical Science

Not all experimental science requires statistics. Many psychological effects are easily repeatable and so can be immediately replicated with no need for statistics. Most of the psychological phenomena of sensation and perception—judging brightness or loudness, most visual illusions—as well as reflex properties, short-term memory, even IQ measures, are unaffected by whatever happens after the first test, so it can be reliably repeated. There is no replication crisis in these areas.[1]

Another area is that of learning and conditioning, especially in animals. Much research on operant conditioning, for example, involves exposing subjects to a reward schedule. In a fixed interval (FI) schedule, for instance, a hungry pigeon or rat gets a bit of food for the first peck or lever press sixty seconds (say) after the last bit of food. No matter what the animal's history, after a few dozen rewards delivered in this way, essentially every subject learns the new pattern. They will wait the given time (proportional to the interfood interval) after each reward, then respond until the next reward. This pattern will recur every time the animal is exposed to the situation,[2] no matter what its experience in between exposures. Many other reward schedules produce reliable behavior patterns like this. No statistics are required: the experiment can be repeated an indefinite number of times with the same subject. The results are replicable, although no one in this area felt the need to think of the field in this way—any more than Faraday worried about the replicability of his repeatable experiments with coils and magnets.

Group-Averaging and the NHST Method

Not all learning experiments can be done with individual subjects, and for good reason. Consider two animals, one (Alpha) trained with procedure *A* followed by *B*, the second (Beta) with *C* followed by *B*. Their behavior under condition *B* we may assume to be identical. But if both animals are now switched to a fourth procedure *X*, they will likely behave very differently because of their different histories, *AB* versus *CB*. In other words, the *behavior* of the two animals in condition *B* may look the same, but their *states* are not. To understand the difference, the two animals must be compared. Since they may well differ in other ways, the natural thing is to compare two groups that differ only in their treatment sequence, hence the ubiquity of the between-group method.

Here is another example: factors affecting speed of learning cannot easily be studied with a single subject, because most learning effects are not reversible—you can't have the subject learning under one set of conditions, "unlearn" him, and then try another. Hence many, perhaps even the majority, of animal-learning experiments are done with groups. These studies are subject to all the strictures that apply to the Null Hypothesis Statistical Test (NHST) method in general.

Unfortunately, the group-testing method has become so ingrained in psychology that it is used even when the same question could have been answered with individual subjects and no statistics at all. There are many examples, going back fifty years or more. Perhaps the most dramatic is a 2011 experiment by psychologist Daryl Bem purporting to show precognition by averaging guesses of a group of people, when this literally incredible faculty should be demonstrable in repeated tries by one "psychic" individual.[3] Here is a more recent example. Benjamin Phillips and his fellow researchers asked a simple question: What is the effect of the type of reward (*reinforcer*) on the behavior of mice working for reinforcement under different kinds of reinforcement schedules?[4] The effects are almost certainly reversible, hence repeatable. That is, if the mouse is given a largish number of trials with reinforcer *A*, followed by the same number of trials with reinforcer *B*, and then repeated, *ABAB*, it is very likely that if there is any real difference between the effects of *A* and *B* it will show up with each repetition. The effect will be replicable.

In fact, the authors used many mice and found and one or two effects. But their method necessarily obscured any individual differences that might exist—even mice may have different tastes, after all—and is subject to the many problems of the NHST method that were pointed out first in a 2005 landmark paper by John Ioannidis provocatively entitled "Why Most Published Research Findings Are False."[5]

An obvious strength of the averaging method is also a weakness. The strength is that a weak or idiosyncratic preference, may be detectable in a "significant" group average. But that is also a weakness, since a small effect should often be ignored. A comparison between two large groups (experimental and control) will often show a small, but significant effect of a new drug, for example. The drug will then be approved by the Food and Drug Administration, allowing the company to market it as more (or at least equally) effective as an older, but perhaps much cheaper, alternative. Good for the company; not so good, probably, for patients. As many people have pointed out, *effect size* is as or more important than statistical significance.[6]

Although I have not done a careful count, it is my impression that in learning psychology there are many more journal papers that use groups of subjects analyzed statistically than studies that use individual animals with no need for statistics. This puzzled me for a while because the effort involved in group studies, which require many animals, seems so much greater than the effort required to study a handful of individuals. The reason, I think, is that the group/ statistical method is *algorithmic* (you can just follow "gold-standard" rules)and one in twenty experiments are pretty much guaranteed to get a positive result (because of the too-generous 5 percent "significance level"). The group method is like what operant conditioners call a *variable-ratio schedule* (also known as a "one-armed bandit"): results are strictly proportional to effort. The NHST method usually can be relied upon to produce more positive results (questionable as many of them now turn out to be) for the same amount of effort, hence more published papers than the single-subject method. It is for this reason much more congenial to the perverse incentives under which most scientists must work. Perhaps that is why this flawed and rather unimaginative method remains so popular.

p-*Hacking*

Reliance on statistical methods has had an unfortunate side effect. Many studies involve not one but several dependent variables, properties of the group that are measured, and several independent variables, things that are manipulated by the experimenter or, for a correlational study, other things that are measured. The problem is endemic in correlational studies. For example, Christie Aschwanden imagines a study to examine whether the U.S. economy is affected by whether Republicans or Democrats are in charge.How does one define "in charge"? The site looks at data from 1948 through 2015 and provides an interactive list of *independent variables*, ranging from presidents and governors to representatives, and of *dependent variables*, including employment, inflation and gross domestic product.[7] This allows for a total of not just 4 x 4 but many more comparisons—three independent variables versus two dependent variables, for example.

There are two problems with such a study: First and foremost, it is just correlational; even if a positive result is found, causation is not proved. And what if there is a delay? A policy implemented now may only affect the economy two or three years later. But second, given a large enough dataset, it is unlikely that *no* correlation will be found. In this case governors versus gross domestic product yields a "significant"—publishable—result on the FiveThirtyEight simulator. Should we conclude that the more Democrat governors the better for the collective wealth? Of course not, but results like this get wide circulation every day.

Another website shows a nice example in which there is a very high correlation between the number of letters in the winning word of the Scripps National Spelling Bee and the number of people killed by venomous spider bites from 1999 to 2009.[8] Most people will not be taken in by the spiders but will swallow political conclusions with little hesitation.

Looking for a "significant" result in this way—trying various comparisons until you get a significant p—is called "p-hacking." But a correlation arrived at this way is of value only as a *hypothesis*. It may be useful as the start of a new study. It is totally illegitimate as a conclusion. If it can be independently tested, which is often difficult or impossible in a correlational study but is perfectly possible in an experimental one, it may turn out to be believable; otherwise it is much more likely to be "noise" rather than "signal," confusing rather than informing.

Thoughtful critics have argued persuasively for *preregistration* of hypotheses as a partial solution to the problem of treating "found" results as confirmed hypotheses.[9] Several scientific organizations have already implemented this idea and provided mild inducements to conformity.[10] There are at least two problems with this level of regulation. Preregistration presumes that scientists must be induced, perhaps even *coerced*, to do their research honestly—not to treat an accidentally significant result as if they began with it as a hypothesis. Such a policy seems to assume either that many scientists are simply stupid (they don't understand the problems with p-hacking) or dishonest (they know but don't care). A researcher who is both competent and honestly desirous of finding the truth simply would not behave in that way. Forcing a dishonest researcher to preregister will just test his ingenuity and challenge him to come up with some other shortcut to quick publication. There is no substitute for honesty in science. In the absence of honesty, imposing or even just recommending, practices like preregistration is likely to produce rigid and uncreative work that is unlikely to advance knowledge and may well retard it. The NHST method already encourages a mechanical approach to science that added regulation will only amplify. A dishonest or dimwitted scientist who must be induced to follow what he should know to be correct practice should find some other line of work.

Correlation versus Causation

Causation is much discussed, especially in disciplines such as economics[11] and epidemiology, where experiment, the most direct and obvious test for causation, is usually impossible. It is a truism that causation cannot be reliably inferred from correlation. The reason is that A may be correlated[12] with B not because A causes B (or vice versa), but because both are caused by (unknown variable) C. Statistician Ronald A. Fisher famously commented: "He [statistician Udny Yule] said that in the years in which a large number of apples were imported into Great Britain, there were also a large number of divorces. The correlation was large, statistically significant at a high level of significance, unmistakable."[13] We don't know what third cause, C, may lie behind this odd correlation, or the spider bites versus word length correlation mentioned earlier. But we can be pretty sure that the apples did not cause the divorces, or vice versa. But when A and B are linked in the popular mind, like obesity and ill-health, or cancer and secondhand smoke, the medical establishment happily lets correlation stand for cause.

The only sure way to prove causation is by experiment. No amount of correlation, of apples with divorces or even smoking and its various supposed sequelae, is as good as actually presenting the cause and getting the effect. And doing this repeatedly.

The Fisherian Method Is Inappropriate for Basic Science

The NHST method, like any experimental method, has its problems. They can usually be overcome. But there is one problem that I think is essentially insuperable and makes NHST simply inappropriate for basic science.

The replication problem is fixable. Various solutions have been proposed. In July 2017 a letter to *Science*, signed by more than seventy statisticians led by Daniel J. Benjamin,[14] suggested that a solution to the NHST-replicability problem is to set the criterion (alpha level) for rejecting the null (no-effect) hypothesis at $p = .005$ rather than

Ronald Fisher's suggestion of $p = .05$, the then-standard.[15] The authors argued that rather than choosing a one in twenty (or less) chance of error as good enough to accept the hypothesis that your treatment has an effect, the standard should be upped to one in two hundred.

Replicability would sometimes be improved by a tougher criterion;[16] but a p-value this small would also eliminate much social-science research that uses NHST; the publication rate in social and biomedical science would take a big hit. Partly for this reason, more than eighty scientists signed a November 2017 letter to *Nature* rejecting the suggestion of Benjamin and his colleagues and instead recommending "that the label 'statistically significant' should no longer be used" and concluding instead "that researchers should transparently report and justify all choices they make when designing a study, including the alpha [critical p-value] level."[17]

The situation seems to have settled down after that point.[18] There have been some efforts at mitigation: preregistering hypotheses, using larger groups, et cetera. But the NHST method continues to be widely used.

The emphasis in both of these long letters is on the replication issue. But a little thought shows that the Fisherian NHST method is in fact *completely inappropriate* for basic science. The reason is depressingly simple.

The NHST method was invented by Ronald Fisher, who was working in an applied environment where a decision *had* to be made—between two fertilizers or other treatments of an agricultural plot, for example.[19] Each fertilizer had a certain estimated net benefit. The one with the significantly higher benefit was then chosen. The cost of an error—choosing the worse fertilizer—is small and potentially measurable. In this case, cost is not an issue. It is only necessary to answer the question, Which is better, probably? For that choice, the 5-percent criterion is perfectly appropriate.

In basic science, the situation is very different: the two choices are "confirm" or "don't know," but *the cost of error is very much higher*. The benefit of seeming to confirm a true experimental hypothesis (that is, rejecting the null hypothesis) is a modest contribution to knowledge. But the cost of an error, seeming to confirm a hypothesis that is in fact false (false positive, Type II error), may be very high, both to science and society.[20] Follow-up studies, in some cases very many studies, will go down the rabbit hole and both waste time and, probably, generate more errors. And as a recent report points out, the cost, both human and financial, of public policies based on a scientific error can be enormous.[21] If the concept of statistical significance is to be retained, the suggestion of the *Science* letter, setting the alpha level very low indeed, should probably be adopted. But the ultimate conclusion is surely closer to the *Nature* letter suggestion that the "truth versus falsity" implication of a statistical criterion simply be abandoned—just give the null-hypothesis statistical model you are using and the probability of your result according to this model.

The Fisherian method is fine for deciding between two types of fertilizer; it is inadequate for deciding between truth and falsehood. It is simply wrong to rely on the NHST method in basic social or biomedical science.

What is the alternative? The simplest possibility is to step back up a bit and restate whatever probabilistic argument leads to your conclusion. For example, suppose you ask one hundred subjects to make a simple choice between A and B. Suppose sixty-five choose option A and twenty-five choose option B, even though A is in fact the rational (utility-maximizing) choice. Can we conclude that the people (in this context) are *rational* or not? Well, if the real probability of rational choice A is close to one, .9, say, then the chance of getting $x = .65$ or higher is very high: .999 or so. Conversely, if the subjects are in fact choosing at random ($x = .5$), the chance of getting a value as high as .65 is low: .00176. So the truth is more likely to be $x = .9$ than $x = .5$. So what is the true probability? Answer, we don't know: the scientific conclusion is that some do and some don't and we do not understand the reason why—in other words, "more research is needed." The real solution to variable data like this is to reduce the variability by uncovering all the historical and contextual causal factors: "no exceptions." Ingenuity is required. It is worth remembering that Hermann Ebbinghaus discovered the basic laws of memory using just one subject, himself, in studies that would not now meet the rote-but-meaningless standards of the NHST community. So not, "Bend it like Beckham!" but "Think like Ebbinghaus!"

The Object of Inquiry: Individual versus the Group[22]

The father of experimental medicine, Claude Bernard, once wrote, "Science does not permit exceptions."[23] But statistics, the NHST method, exists because of exceptions. If an experimental treatment gave the same, or at least a similar, result in every case, statistics would not be necessary. But going from the group, which is the subject matter for statistics, to the individual, which is the usual research objective of psychology and biomedicine, poses problems that are frequently ignored.

A couple of examples may help. In the first case, the subject of the research really is the group; in the second, the real subject is individual human beings, but the group method is used.

Polling uses a sample to get an estimate of the preferences of the whole population. Let's say that our (random) sample shows 30 percent favoring option A and 70 percent favoring B. How reliable are these numbers as estimates of the population as a whole? If the population is small, the answer depends on the size of the sample in relation to the size of the population. If the population is little larger than the sample, the sample is likely to give an accurate measure of the state of the population.

If the population is large this method is not possible. To get an estimate of the population from a sample that is a small fraction, some kind of *model* is needed. An obvious possibility in this simple case is to compare sampling with unbiased coin tossing. Hypothesis: People are indifferent between *A* and *B*, so the expected outcome of our questionnaire is 50:50. Result (data): 30:70. Test: How likely is a 30:70 result if the population is really 50:50? Answer: With a sample of 100, very low indeed (less than .00008). Conclusion: The population is not indifferent between *A* and *B*. So, we can conclude there is a bias, but we cannot be sure how big it is.[24] On the other hand, the model allows us to compute how probable the data are for different hypothetical biases. For example, if the true bias is not 30:70, but 40:60, the chance of a sample deviation of 30:70 or more is .024, which is not negligible. (Or is it? Like many statistical decisions, this one is rather arbitrary.)

Whether the population is large or small, the aim is to draw conclusions not about the individual decision makers, but about the population as a whole. The method does not violate Bernard's maxim. Since the conclusion is about the sample, there are no exceptions.

Prospect Theory

Here is a contrary example. Experiments in psychology and decision theory are aimed at understanding not the population as a whole, but the individual. Very often the method they use, or at least the conclusion they draw, is inappropriate. For example, in 1979 Daniel Kahneman and Amos Tversky came up with a clever way to study human choice behavior. This work eventually led to many thousands of citations and an economics Nobel prize in 2002. Kahneman and Tversky consulted their own intuitions and came up with simple problems that they then asked individual subjects to solve. The results, statistically significant according to the standards of the time, generally confirmed their intuition. They replicated many results in different countries.

Their classic paper describes how subjects were asked to pick one of two choices. In one case, the choices were A: 4000 (Israeli currency) with probability .2, or B: 3000 with probability .25. 65 percent of the subjects picked A, with an expected gain[25] of 800 over B, 35 percent, with an expected gain of 750. This represents rational choice on the part of 65 percent of choosers.

Kahneman and Tversky contrast this result with an apparently similar problem, A: 4000, $p = .8$, versus B: 3000, $p = 1.0$. In this case, 80 percent of subjects chose B, the certain option with the lower expected value (3000 versus 3200), an irrational choice. They then used this contrast, and the results of many similar problems, to come up with an alternative to standard utility theory.[26]

Prospect theory, as Kahneman and Tversky called it, is not like most scientific theories: namely a modest set of assumptions from which many predictions can be deduced. It is rather list of effects—such as the certainty effect (this case), loss-aversion, confirmation bias, and the like. (Wikipedia lists more than one hundred "biases" like this.[27]) Prospect theory attempted, with only partial success, to account for these discrepancies by proposing a modified utility function, but basically it is a set of labels, rather than a theory.[28]

Most importantly for the present argument, prospect theory—any theory that attempts to account for the majority choice—cannot apply to the minority choice. In my example, neither the 35 percent who chose rationally in the first case nor the 20 percent who chose irrationally in the second. If the object of inquiry is the psychology of individual human beings, the work is incomplete. To finish the job, the experimenters need to find out what it takes to produce the same result for all subjects.

Kahneman and Tversky's results and the theory that describes them are a property of groups of people under very restricted conditions. They are not, as is sometimes assumed, general properties of human nature. The method is just opinion polling of an arbitrary population, albeit with a clever set of questions.[29] Yet each effect is presented as a property of human choice behavior *tout court*. Because the group effect is reliable, individual exceptions are ignored. There is little doubt, for example, that subjects given a lesson or two in probability, or faced with a question phrased slightly differently (for example, not "What is your choice?" but "What would a statistician choose?"), would choose differently. Indeed, a few such possibilities have been explored. Under some conditions, a group might even choose differently than an individual. For example, suppose the subjects were allowed to consult among themselves before choosing. If the choice were between, say, $1000 for sure versus a .6 chance of $2000, would it not be wise for the group of, say, twenty subjects to agree amongst themselves to always choose the .6-and-$2000 option, and then split the proceeds evenly—thus ensuring that everyone would probably get more than $1000?

People's choices in a problem like this almost certainly depend on the absolute quantities involved. Who can doubt that if subjects were offered a credible choice between, say, $10 million for sure versus a .6 chance of $20 million, close to 100 percent would choose the sure thing?[30] Until the theory is developed to account for the minority choice and the effect of absolute amount, until it is in a form that allows no individual exceptions, it cannot pretend to be an adequate model of human choice behavior. The same reservations apply to much of the spin-off field of behavioral economics.

Despite its inapplicability to individual behavior, prospect theory is usually presented as an accurate picture of human choice. Kahneman and Tversky were not alone in treating their group theory as a theory of individual choice behavior. The field of NHST made and continues to make the same mistake: a significant group result is treated as a property of people in general.

The aim of these choice experiments, unlike the polling examples I gave earlier, is to understand not groups but individuals. Even if a study is perfectly reproducible, to go beyond a group average, the experimenters would need to look at the causal factors that differentiate individuals who respond differently. What is it, about the constitution or personal histories of individuals, that makes them respond differently to the same question? Solving this problem, satisfying Claude Bernard's admirable axiom, is obviously much tougher than simply asking, "Do you prefer $3000 for sure or a .8 chance of $4000?" But until this problem is solved, both prospect theory and expected-utility theory—not to mention numberless other psychological theories—must give an incomplete, not to say distorted, picture of individual human psychology.

Adoption of the Fisherian method has fueled a steady growth in the numbers of papers in psychology and, yes, many of these have turned out to be non-replicable. But this may not be the most serious problem. Even replicable group studies are uninterpretable as to the behavior of individuals. And because they are uninterpretable, the parts of the field that have adopted these methods have stalled. No testable integrative theories have emerged. More and more papers are produced, but the field does not cumulate, as those parts of psychology that use the less popular and more challenging single-subject method have done and continue to do. As far as our scientific understanding of individual psychology is concerned, the Fisherian method seems to be a dead-end.

Acknowledgements

M any people helped, directly or indirectly with this book. I thank Peter Morcombe, Peder Zane, John Malone, Peter Arcidiacono, Judith Curry, William Happer, Prasad Kasibhatla, John Droz, Jeffrey Tucker, and a very helpful anonymous reviewer; Clive Wynne, Jay Schalin, Marc Branch, James Taranto, Katy Delay, Toby Young, the HOPE group at Duke, Josefina Tiryakian, George Leef, James Enstrom, John Horgan, Armand Leroi, David Hockney, Michael Munger, Drake Jacobs, Oscar Barnes, and Sumantra Maitra. I especially thank Alan Silberberg for his thoughtful comments on the book. I am also grateful to the journal *Academic Questions*, the Martin Center website, and the website Quillette, as well as my own blog on *Psychology Today*, which published earlier ruminations on many of the issues discussed in this book. Any errors that remain are of course entirely my fault.

Notes

Preface

1. Sociologist Max Weber made a very similar distinction between *facts* and *values*. See, for example, Steve Hoenisch, "Max Weber's View of Objectivity in Social Science," https://criticism.com/md/weber1.html.
2. Lily Levin, "It Is All Racism," *The Chronicle*, January 10, 2021, https://www.dukechronicle.com/article/2021/01/it-is-all-racism-duke-magazine.
3. Alvin Weinberg, "Science and Trans-Science," *Minerva* 10 (April 1972): 209–222.
4. Brooke Singman and Ronn Blitzer, "Biden Calls Climate Crisis 'an Existential Threat,' Apologizes for Trump Pulling Out of Paris Accord," Fox News, November 1, 2021, https://www.foxnews.com/politics/biden-calls-climate-crisis-an-existential-threat-to-human-existence-as-we-know-it-at-cop26.

Chapter 1: Has Secular Humanism Made Science a Religion?

1. Wikipedia, s.v. "Peloza v. Capistrano School District," https://en.wikipedia.org/wiki/Peloza_v._Capistrano_School_District.
2. Rod Liddle, *The Trouble with Atheism (Religious Documentary)*, Real Stories, July 10, 2018, https://www.youtube.com/watch?v=g6MrktRKfJU.
3. A case made forcefully by philosopher John Gray, "The Closed Mind of Richard Dawkins: His Atheism Is Its Own Kind of Narrow Religion," *New Statesman*, October 4, 2014, https://www.newstatesman.com/culture/2014/10/closed-mind-richard-dawkins.

4. A similar point was made many by the philosopher Erasmus (1466–1536), who distinguished between mandatory doctrines and nonobligatory *adiaphora*, which "were to be distinguished from godly directives, like the Decalogue and the precepts of Christ and his Apostles." Gary Remer, *Humanism and the Rhetoric of Toleration* (University Park: Pennsylvania State University Press, 1996), 50.

5. "On a strict understanding [of David Hume] . . . 'true religion' is simply ethical conduct . . ." Paul Russell, "'True Religion' and Hume's Practical Atheism," in *Sceptical Doubt and Disbelief in Modern European Thought: A New Pan-American Dialogue*, ed. Vicente Raga Rosaleny and Plínio Junqueira Smith (Dordrecht: Springer, 2021).

6. "Attitudes on Same-Sex Marriage: Public Opinion on Same-Sex Marriage," Pew Research Center, March 14, 2019, https://www.pewforum.org/fact-sheet/changing-attitudes-on-gay-marriage/.

7. Bill Chappell, "Supreme Court Declares Same-Sex Marriage Legal in All 50 States," NPR, June 26, 2015, https://www.npr.org/sections/thetwo-way/2015/06/26/417717613/supreme-court-rules-all-states-must-allow-same-sex-marriages.

8. Question: "I hear that Jim just got married: What's her name?" Answer: "Sebastian."

9. The Harm Principle, defined in David Brink, "Mill's Moral and Political Philosophy," *The Stanford Encyclopedia of Philosophy* (Spring 2022 Edition), ed. Edward N. Zalta, https://plato.stanford.edu/entries/mill-moral-political/#HarPri, shares the defect of utilitarianism/consequentialism in general. Even if we know what is good and bad, we are still more or less ignorant of the effects, especially delayed or distributed effects, of any apparently benign policy. Consider abortion: in terms of personal autonomy, "control of our bodies," and so on, freely available pregnancy termination is obviously good. Religiously based opposition is dismissed as "superstition." But we don't know what the effect of the normalization of the practice on society as a whole will be. Will increased availability of abortion promote or discourage stable marriage? Or will a freer, better, social structure than the traditional nuclear family arise as a consequence? Or, as many tend to assume, will the practice have no effect beyond the potential mother?

10. Saying the N-word while black is of course permitted: see, for example, the movie *Concrete Cowboy* and almost any rap song.

11. Colleen Flaherty, "Too Taboo for Class?" Inside Higher Education, February 1, 2019, https://www.insidehighered.com/news/2019/02/01/professor-suspended-using-n-word-class-discussion-language-james-baldwin-essay.

12. Raymond Hernandez, "Richard Blumenthal's Words on Vietnam Service Differ from History," *New York Times*, May 17, 2010.

13. John Staddon, "Microaggression, *Mens Rea* and the Unconscious Mind: Are We Unconciously Demeaning Protected Groups," *Psychology Today*, May 12, 2016, https://www.psychologytoday.com/us/blog/adaptive-behavior/201605/microaggression-mens-rea-and-the-unconscious-mind; John Staddon, "It's All About Power: Microaggression Does Exist, but Not Where You Thought," *Psychology Today*, June 1, 2016, https://www.psychologytoday.com/us/blog/adaptive-behavior/201606/its-all-about-power.

14. Adam Liptak, "Supreme Court Allows 40-Foot Peace Cross on State Property," *New York Times,* June 20, 2019.

15. Adam Liptak, "40-Foot Cross Divides a Community and Prompts a Supreme Court Battle," *New York Times*, February 24, 2019.

16. Joy Pullman, "Left to State Supreme Court Candidate: You Can't Be a Good Judge Because You're a Christian," *The Federalist*, February 18, 2019, https://thefederalist.com/2019/02/18/leftists-wi-supreme-court-candidate-cant-good-judge-youre-christian/.

17. This argument, which seemed obvious to me, nevertheless elicited an irate response from evolutionary biologist Jerry Coyne, who blogged: "The worst article ever to appear in Quillette: Psychologist declares secular humanism a 'religion,'" Why Evolution Is True, April 12, 2019, https://whyevolutionistrue.com/2019/04/12/the-worse-article-ever-to-appear-in-quillette-psychologist-declares-secular-humanism-a-religion/. An online exchange of views followed. See John Staddon, "Values, Even Secular Ones Depend on Faith: A Reply to Jerry Coyne," Quillette, April 28, 2019, https://quillette.com/2019/04/28/values-even-secular-ones-depend-on-faith-a-reply-to-jerry-coyne/; Doug Drake, "Is Secular Humanism a Religion? Is Secular Humanist Morality Really Subjective?," Helian Unbound, April 21, 2019, https://helian.net/blog/2019/04/21/morality/is-secular-humanism-a-religion-is-secular-humanist-morality-really-subjective/.

Chapter 2: Science and Faith: Can Morality Be Deduced from the Facts of Science?

1. Edward O. Wilson, "The Biological Basis of Morality," *The Atlantic*, April 1998, https://www.theatlantic.com/magazine/archive/1998/04/

the-biological-basis-of-morality/377087/. The quote also appears in Wilson's book *Consilience: The Unity of Knowledge* (New York: Alfred Knopf, 1998), 274, although he always vehemently denied falling for the naturalistic fallacy; see, for example, Alice Dreger, "A Conversation with E. O. Wilson (1929–2021)," Quillette, December 29, 2021, https://quillette.com/2021/12/29/speaking-with-e-o-wilson/.

2. Tom Junod, "E. O. Wilson: What I've Learned," *Esquire*, January 5, 2009, http://www.esquire.com/features/what-ivelearned/eo-wilson-quotes-0109.

3. John Staddon, "Scientific Imperialism and Behaviorist Epistemology," *Behavior and Philosophy* 32, no. 1 (2004): 231–42.

4. James E. Taylor, "The New Atheists," in *Internet Encyclopedia of Philosophy*, https://iep.utm.edu/n-atheis/#:~:text=The%20New%20Atheists%20are%20authors,Daniel%20Dennett%2C%20and%20Christopher%20Hitchens.

5. A quote that once appeared on The Humanist website but has since expired. https://thehumanist.com/humanist/articles/dawkins.html.

6. Richard Dawkins, *The God Delusion* (London: Bantam Press, 2006), 308.

7. Sam Harris, *The Moral Landscape: How Science Can Determine Human Values* (New York: Free Press, 2010).

8. Sam Harris, "Science Can Answer Moral Questions," TED Talk, February 2010, https://www.ted.com/talks/sam_harris_science_can_answer_moral_questions.

9. Harris, *The Moral Landscape*.

10. Harris, "Science Can Answer Moral Questions."

11. Charles Darwin, *On the Origin of the Species by Means of Natural Selection, or the Preservation of Favoured Races in the Struggle for Life* (London: 1859), 68, http://darwin-online.org.uk/Variorum/1859/1859-68-dns.html.

12. Leon Zitzer, *Darwin's Racism: The Definitive Case, Along with a Close Look at Some of the Forgotten, Genuine Humanitarians of That Time* (Bloomington, Indiana: iUniverse, 2016), http://onala.free.fr/zitzer2016.pdf. This is a summary of a much longer book accusing Darwin of racism. On the contrary, authors Adrian Desmond and James Moore, *Darwin's Sacred Cause: Race, Slavery, and the Quest for Human Origins* (London: Allen Lane, 2009), trace Darwin's whole evolutionary project to his hatred of slavery . . . go figure.

13. David Hume, *David Hume Collection: A Treatise of Human Nature*, (1738–40), Kindle.

14. Ibid.

15. Harris, *The Moral Landscape*.

16. It is interesting that some scholars trace the decline of Islamic science, which flourished from the seventh to the fifteenth centuries, to the rise of the dogmatic religious movement, the Ash'arites: "At the heart of Ash'ari metaphysics is the idea of occasionalism, a doctrine that denies natural causality." Hillel Ofek, "Why the Arabic World Turned Away from Science: On the Lost Golden Age and the Rejection of Reason," *New Atlantis*, Winter 2011, https://www.thenewatlantis.com/publications/why-the-arabic-world-turned-away-from-science. See also Stanley L. Jaki, *The Savior of Science* (Washington, D.C.: Regnery Gateway, 1988), which argues that Christianity provides the faith, necessary for science, that nature is a stable order. Nicholas Wade in *A Troublesome Inheritance: Genes, Race and Human History* (New York: Penguin, 2014) has made a similar point. Atheist and theoretical physicist Jim Al-Khalili, on the other hand, argues persuasively that the eventual decline of Arab/Islamic science was not due to the Islamic faith but to the fall of its various supporting empires. J. Al-Khalili, *Pathfinders: The Golden Age of Arabic Science* (London: Penguin, 2010).See also Tim Radford, "Pathfinders: The Golden Age of Arabic Science by Jim al-Khalili—Review," https://www.theguardian.com/books/2010/oct/23/arabic-science-jim-alkhalili-review.

17. Daniel C. Dennett, quoted in Harris, *The Moral Landscape.*

Chapter 3: Science and Faith: Darwin to the Rescue?

1. Michael Bradie and William Harms, "Evolutionary Epistemology," *The Stanford Encyclopedia of Philosophy*, ed. Edward N. Zalta, Spring 2020, https://plato.stanford.edu/archives/spr2020/entries/epistemology-evolutionary/.

2. Sean Carroll, "You Can't Derive Ought from Is," *Discover*, May 3, 2010, https://www.discovermagazine.com/mind/you-cant-derive-ought-from-is.

3. B. F. Skinner, *Beyond Freedom and Dignity* (New York: Alfred Knopf, 1971), 102.

4. G. E. Moore, *Principia Ethica* (Cambridge: Cambridge University Press, 1903), https://en.wikipedia.org/wiki/Principia_Ethica#/media/File:Principia_Ethica_title_page.png.

5. E. O. Wilson, *Consilience: The Unity of Knowledge* (New York: Alfred Knopf, 1998), 262. Nevertheless, by taking any account of genetics, even after his death, Wilson is routinely labeled a "racist" in what used to be respectable journals, such as *Scientific American*: Monica R. McLemore, "The Complicated Legacy of E. O. Wilson: We Must Reckon with His and Other

Scientists' Racist Ideas if We Want an Equitable Future," *Scientific American*, December 29, 2021, https://www.scientificamerican.com/article/the-complicated-legacy-of-e-o-wilson/.

6. Skinner, *Beyond Freedom and Dignity*, 164.
7. Karl R. Popper, *The Open Society and Its Enemies* (Princeton, New Jersey: Princeton University Press, 1950), 665. See also Popper's *Conjectures and Refutations: The Growth of Scientific Knowledge* (New York: Basic Books, 1962).
8. Amanda Macias, "America's Oldest Veteran, Known for Smoking Cigars and Drinking Whiskey, Dies at Age 112," CNBC, December 28, 2018, https://www.cnbc.com/2018/12/28/richard-overton-dies-at-the-age-of-112.html.
9. Stephen C. Stearns, *The Evolution of Life Histories* (Oxford: Oxford University Press, 1992).
10. Victor Davis Hanson, "What Millenials Can Learn from the Greatest Generation," *Newsweek*, January 1, 2018, https://www.newsweek.com/what-millennials-can-learn-greatest-generation-766296.
11. See review in John Staddon, *Unlucky Strike: Private Health and the Science, Law and Politics of Smoking* (Buckingham: University of Buckingham Press, 2014); text is available at https://dukespace.lib.duke.edu/dspace/handle/10161/22366.
12. C. K. Chan, "Eugenics on the Rise: A Report from Singapore," *International Journal of Health Services* 15, no. 4 (1985): 707–712.
13. Russian anarchist Peter Kropotkin (1842–1921) thought bees a fine model for a human utopia: *Mutual Aid: A Factor of Evolution*, (New York: McClure Phillips & Co., 1902); text is available at https://www.marxists.org/reference/archive/kropotkin-peter/1902/mutual-aid/ch01.htm.
14. James Q. Wilson, "Democracy for All?," *Commentary*, March 2000, https://www.commentarymagazine.com/articles/james-wilson/democracy-for-all/.

Chapter 4: Was Darwin Wrong or Just Misunderstood?

1. Christopher Booker, *Global Warming: A Case Study in Groupthink; How Science Can Shed New Light on the Most Important 'Non-Debate' of Our Time* (London: The Global Warming Policy Foundation, 2018), https://www.thegwpf.org/content/uploads/2018/02/Groupthink.pdf; Christopher Booker, "All Done with Passive Smoke and Mirrors," *The Telegraph*, July 1, 2007, https://www.telegraph.co.uk/news/1556118/Christopher-Bookers-notebook.html.

2. Christopher Booker, "Scientists in Hiding," *The Spectator*, September 18, 2010, https://www.spectator.co.uk/article/scientists-in-hiding.

3. Peter R. Grant and B. Rosemary Grant, "Evolution of Character Displacement in Darwin's Finches," *Science* 313, no. 5784 (July 14, 2006): 224–26, https://www.science.org/doi/10.1126/science.1128374.

4. Wikipedia, s.v. "Peppered Moth Evolution," https://en.wikipedia.org/wiki/Peppered_moth_evolution.

5. Charles Darwin, *The Structure and Distribution of Coral Reefs* (New York: Appleton, 1914), https://www.google.com/books/edition/The_Structure_and_Distribution_of_Coral/-X5GAQAAMAAJ?hl=en&gbpv=1&printsec=frontcover.

6. "The ants carrying the cocoons did not appear to be emigrating. . . . But when I looked closely I found that all the cocoons were empty cases. . . . Now here I think we have one instinct in contest with another and mistaken one. The first instinct being to carry the empty cocoons out of the nest. . . . And then came in the contest with the other very powerful instinct of preserving and carrying their cocoons as long as possible; and this they could not help doing although the cocoons were empty. According as the one or other instinct was the stronger in each individual ant, so did it carry the empty cocoon to a greater or less distance." Charles Darwin, *Life and Letters*, ed. Francis Darwin, vol. 3 (London: John Murray, 1887), 192, http://darwin-online.org.uk/content/frameset?pageseq=1&itemID=F1452.3&viewtype=text.

7. Steve Connor, "The Original Theory of Evolution . . . Were It Not for the Farmer Who Came Up with It, 60 Years before Darwin," *The Independent*, October 16, 2003, https://www.independent.co.uk/news/science/original-theory-evolution-were-it-not-farmer-who-came-it-60-years-darwin-91580.html.

8. Richard Dawkins, *The Genius of Charles Darwin—Part 1: Life, Darwin & Everything*, YouTube, July 5, 2013, https://www.youtube.com/watch?v=ZtkZMAmgHaU.

9. Ibid. The same metaphor was used by B. F. Skinner when he discussed the shaping of behavior via operant conditioning before he later put the same idea in a Darwinian context. Gail B. Peterson, "The Discovery of Shaping: B. F. Skinner's Big Surprise," Cambridge Center for Behavioral Studies, Inc., http://www.behavior.org/resources/453.pdf.

10. Charles Darwin, *On the Origin of the Species by Means of Natural Selection, or the Preservation of Favoured Races in the Struggle for Life* (London: 1872), 189, http://darwin-online.org.uk/Variorum/1872/1872-146-dns.html.

11. "Evolution of the Eye," from *Evolution: "Darwin's Dangerous Idea,"* hosted by Dan-Erik Nilsson on PBS, WGBH Education Foundation and Clear Blue Sky Productions, Inc., 2001, http://www.pbs.org/wgbh/evolution/library/01/1/l_011_01.html.

12. See, for example, Stefan Wawersik and Richard L. Maas, "Vertebrate Eye Development as Modeled in *Drosophila,*" *Human Molecular Genetics* 9, no. 6 (2000): 917–25, https://academic.oup.com/hmg/article/9/6/917/618651.

13. Charles Darwin, *The Descent of Man, and Selection in Relation to Sex* (London: 1871); Wikipedia, s.v. *"The Descent of Man, and Selection in Relation to Sex,"* https://en.wikipedia.org/wiki/The_Descent_of_Man,_and_Selection_in_Relation_to_Sex; Patricia L. R. Brennan, "Sexual Selection," *Nature Education Knowledge* 3, no. 10 (2010): 79, https://www.nature.com/scitable/knowledge/library/sexual-selection-13255240/.

14. The doctrine that facts about the past (in geology, for example) should be explained by processes visible in the present. Wikipedia, s.v. "Uniformitarianism," https://en.wikipedia.org/wiki/Uniformitarianism.

15. Darwin, *Origin of the Species*, 8, http://darwin-online.org.uk/Variorum/1872/1872-8-dns.html.

16. Ibid., 9, http://darwin-online.org.uk/Variorum/1872/1872-9-dns.html.

17. See Lee Alan Dugatkin, "The Silver Fox Domestication Experiment," *Evolution: Education and Outreach* 11, no. 16 (2018), https://evolution-outreach.biomedcentral.com/articles/10.1186/s12052-018-0090-x.

18. Michael R. Dietrich, "From Hopeful Monsters to Homeotic Effects: Richard Goldschmidt's Integration of Development, Evolution, and Genetics," *American Zoologist* 40, no. 5 (October 2000): 738–47, https://academic.oup.com/icb/article/40/5/738/157070.

19. Douglas Fox, "What Sparked the Cambrian Explosion?," *Nature* 530 (February 2016): 268–70, https://www.nature.com/news/what-sparked-the-cambrian-explosion-1.19379.

20. W. M. Fitch and F. J. Ayala, "Tempo and Mode in Evolution," colloquium paper printed in *Proceedings of the National Academy of Sciences of the United States of America* 91, no. 15 (July 1994): 6717–20.

21. Stephen Jay Gould and Niles Eldredge, "Punctuated Equilibria: The Tempo and Mode of Evolution Reconsidered," *Paleobiology* 3, no. 2 (Spring 1977): 115–51.

22. Gabriel Dover, "Molecular Drive," *Trends in Genetics* 18, no. 11 (November 2002): 587–89.

23. Gabriel Dover, "Molecular Drive: A Cohesive Mode of Species Evolution," *Nature* 299, no. 5879 (September 1982): 111–17.
24. Wikipedia, s.v. "Genetic Hitchhiking," https://en.wikipedia.org/wiki/Genetic_hitchhiking.
25. Wikipedia, s.v. "Genetic Drift," https://en.wikipedia.org/wiki/Genetic_drift.
26. D'Arcy Wentworth Thompson, *On Growth and Form* (Cambridge: Cambridge University Press, 1917).
27. Philip Ball, "In Retrospect: On Growth and Form," *Nature* 494 (February 2013): 3233, https://www.nature.com/articles/494032a.
28. Discussed by Armand Leroi in "What Darwin Didn't Know," BBC Four, YouTube, March 23, 2013, https://www.youtube.com/watch?v=cPdM musV1ZU. This is perhaps the best video account of the history and development of Darwinism.
29. The point I am making here is different from *orthogenesis,* a somewhat controversial term, usually taken to mean that evolution is goal directed. The constraints I discuss say nothing about a goal, only that the options for future evolution from any starting point are constrained in ways yet to be fully defined.
30. Randy J. Guliuzza, "Major Evolutionary Blunders: Our Useful Appendix— Evidence of Design, Not Evolution," Institute for Creation Research, January 29, 2016, https://www.icr.org/article/major-evolutionary-blunders-our-useful.

Chapter 5: Are We Losing Our Way?

1. Gina Kolata, "So Many Research Scientists, So Few Openings as Professors," *New York Times,* July 14, 2016.
2. Paula Stephan, "Too Many Scientists?," *Chemistry World*, January 22, 2013, https://www.chemistryworld.com/opinion/too-many-scientists/5820.article.
3. "The Disposable Academic: Why Doing a PhD Is Often a Waste of Time," *The Economist,* December 16, 2010.
4. Dante Ramos, "Adjunct Professors Unionize, Revealing Deeper Malaise in Higher Ed," *Boston Globe*, March 24, 2016.
5. Yi Xue and Richard C. Larson, "STEM Crisis or STEM Surplus? Yes and Yes," *Monthly Labor Review*, May 26, 2018, https://pubmed.ncbi.nlm.nih.gov/29422698/. STEM = Science, Technology, Engineering, and Math.
6. Kolata, "So Many Research Scientists." See also Richard C. Larson, Navid Ghaffarzadegan, and Yi Xue, "Too Many PhD Graduates or Too Few Academic Job Openings: The Basic Reproductive Number R_0 in Academia,"

Systems Research and Behavioral Science 31, no. 6 (November–December 2013): 745–50.

7. Vannevar Bush, *Science, the Endless Frontier: A Report to the President by Vannevar Bush, Director of the Office of Scientific Research and Development, July 1945,* (Washington, D.C.: U.S. Government Printing Office, 1945), chapter 1.

8. Daniel Sarewitz, "Saving Science: Science Isn't Self-Correcting, It's Self-Destructing," *New Atlantis,* Spring/Summer 2016, https://www.thenewatlantis.com/publications/saving-science.

9. Wikipedia, s.v. "Replication Crisis," https://en.wikipedia.org/wiki/Replication_crisis.

10. Paul Romer, "Trouble with Macroeconomics, Update," September 21, 2016, https://paulromer.net/trouble-with-macroeconomics-update/; John Staddon, *Scientific Method: How Science Works, Fails to Work or Pretends to Work,* (London: Taylor and Francis, 2017), chapters 5–7.

11. The implicit association test is one example: Jesse Singal, "The Creators of the Implicit Association Test Should Get Their Story Straight," *New York,* December 4, 2017; Tom Bartlett, "Can We Really Measure Implicit Bias? Maybe Not," *Chronicle of Higher Education,* January 5, 2017, https://www.chronicle.com/article/can-we-really-measure-implicit-bias-maybe-not/. Singal quotes Hart Blanton et al., who observe that "the IAT provides little insight into who will discriminate against whom, and provides no more insight than explicit measures of bias." A neglected competitor for the IAT is satirist Peter Simple's prejudometer "which simply by being pointed at any person could calculate degrees of racism to the nearest prejudon, 'the internationally recognized scientific unit of racial prejudice.'" Mark Steyn, "The Last Edwardian: Michael Wharton (1913–2006), *The Atlantic* (April 2006), https://www.theatlantic.com/magazine/archive/2006/04/the-last-edwardian/304734/; see also Peter Simple, "Boon," *The Telegraph,* April 13, 2001, https://www.telegraph.co.uk/comment/4261218/The-Peter-Simple-Column.html.

12. Algis Valiunas, "The Evangelist of Molecular Biology," *New Atlantis* (Summer/Fall 2017), https://www.thenewatlantis.com/publications/the-evangelist-of-molecular-biology.

13. Robin McKie, "The Ant King's Latest Mission," *The Guardian,* September 30, 2006, https://www.theguardian.com/world/2006/oct/01/usa.science.

14. Null Hypothesis Significance Testing, a method devised by R. A. Fisher for comparing variable data from an experimental and a control group; more in the appendix.
15. David Colquhoun, "An Investigation of the False Discovery Rate and the Misinterpretation of P-Values," *Royal Society Open Science* 1, no. 3 (November 2014), https://doi.org/10.1098/rsos.140216.
16. Angelika Batta, Bhupinder Singh Kalra, Raj Khirasaria, "Trends in FDA Drug Approvals over Last 2 Decades: An Observational Study," *Journal of Family Medicine and Primary Care* 9, no. 1 (January 2020): 105–14.
17. Heather Mac Donald, "How Identity Politics Is Harming the Sciences," *Jewish World Review*, May 15, 2018, https://www.jewishworldreview.com/0518/mac_donald051518.php3. For a summary of the NIH commitment, see Marie A. Bernard and Mike Lauer, "Reaffirming NIH's Commitment to Workforce Diversity, National Institutes of Health, November 3, 2021, https://nexus.od.nih.gov/all/2021/11/03/reaffirming-nihs-commitment-to-workforce-diversity/.
18. "Agency Mission," National Science Foundation, https://www.usaspending.gov/agency/national-science-foundation?fy=2022.
19. Heather Mac Donald, "Even Amid a Pandemic, Federal Science Agencies Continue to Fund Anti-Scientific Diversity Initiatives," *Jewish World Review*, May 11, 2020, https://jewishworldreview.com/0520/mac_donald051120.php3.
20. John Staddon, "The Diversity Dilemma," *Academic Questions* 34, no. 3 (Fall 2021): 109–11, https://www.nas.org/academic-questions/34/3/the-diversity-dilemma.
21. "Racial Equity in STEM Education (EHR Racial Equity)," National Science Foundation, https://beta.nsf.gov/funding/opportunities/racial-equity-stem-education-ehr-racial-equity.
22. "NSF INCLUDES Announces New Alliances Focused on Increasing Equity and Broadening Participation in STEM," National Science Foundation, August 3, 2021, https://www.nsf.gov/news/special_reports/announcements/080321.jsp. INCLUDES (Inclusion across the Nation of Communities of Learners of Underrepresented Discoverers in Engineering and Science) easily wins the prize for most stomach-churning acronym of the decade. The Defense Department will have to look to its laurels! The INCLUDES program began in 2017.
23. "NSF INCLUDES Alliance: The Alliance for Identity-Inclusive Computing Education (AIICE): A Collective Impact Approach to Broadening Participation in Computing," National Science Foundation, August 18, 2021,

https://www.nsf.gov/awardsearch/showAward?AWD_ID=2118453&Historical Awards=false.

24. Wikipedia, s.v. "Asilomar Conference on Recombinamt DNA," https:// en.wikipedia.org/wiki/Asilomar_Conference_on_Recombinant_DNA.

25. "FY 2022 Budget Request to Congress," National Science Foundation, https:// www.nsf.gov/about/budget/fy2022/pdf/11_fy2022.pdf. The 2022 NSF budget request under the heading "Broadening Participation Programs" is $1.4 billion. See also "FY22 Budget Request: National Science Foundation," American Institute of Physics, June 3, 2021, https://www.aip.org/fyi/2021/ fy22-budget-request-national-science-foundation, which states under the heading "Demographic diversity" that "total funding for programs explicitly focused on diversifying the STEM workforce would increase by 26% to $582 million under the [2022] request."

Chapter 6: Scientific Publishing

1. Daniel Dennett, "Postmodernism and Truth" (paper, World Congress of Philosophy, August 13, 1998), https://ase.tufts.edu/cogstud/dennett/papers/ postmod.tru.htm#N_1_.

2. Cost figures valid as of 2018.

3. "How Many Science Journals," written by hbasset, Science Intelligence and InfoPros, January 23, 2012, https://scienceintelligence.wordpress. com/2012/01/23/how-many-science-journals/.

4. Wikipedia, s.v. "List of Psychology Journals," https://en.wikipedia.org/wiki/ List_of_psychology_journals.

5. Michael Mills, "Plan S—What Is Its Meaning for Open Access Journals and for the JACMP?," *Journal of Applied Clinical Medical Physics* 20, no. 3 (March 2019): 4–6, https://www.ncbi.nlm.nih.gov/pmc/articles/PMC6414135/.

6. "Part 1: The Plan S Principles," Plan S, https://www.coalition-s.org/ addendum-to-the-coalition-s-guidance-on-the-implementation-of-plan-s/ principles-and-implementation/.

7. "Organizations Endorsing Plan S and Working Jointly on Its Implementation," Plans S, https://www.coalition-s.org/organisations/.

8. Martin Enserink, "European Science Funders Ban Grantees from Publishing in Paywalled Journals," *Science*, September 4, 2018, https://www.sciencemag. org/news/2018/09/european-science-funders-ban-grantees-publishing- paywalled-journals.

9. "Appendix 138: Supplementary Evidence from Nature Publishing Group," (report prepared by Select Committee on Science and Technology, UK

Parliament, July 20, 2004), https://publications.parliament.uk/pa/cm200304/cmselect/cmsctech/399/399we163.htm.

10. "Open Access Scientific Publishing," *The Guardian*, https://www.theguardian.com/science/open-access-scientific-publishing.

11. Philip Mirowki, "Hell Is Truth Seen Too Late"(paper, *boundary 2*, July 2017), https://www.ineteconomics.org/uploads/papers/Mirowski-Hell-is-Truth-Seen-Too-Late.pdf. Economist Philip Mirowski paints a dire picture for the future of "open science," seeing great dangers in a "neo-liberal," market-driven takeover of science. He shows a patent application, by Elsevier, for an automated peer-review process that is alarming in several ways. But he leaves out of account many of the factors that conduce to scientific discovery. The long article (the science bit is toward the end) is well worth reading.

12. Alex Hern and Pamela Duncan, "Predatory Publishers: The Journals That Turn Out Fake Science," *The Guardian*, August 10, 2018, https://www.theguardian.com/technology/2018/aug/10/predatory-publishers-the-journals-who-churn-out-fake-science.

13. George Monbiot, "Academic Publishers Make Murdoch Look like a Socialist," *The Guardian*, August 29, 2011, https://www.theguardian.com/commentisfree/2011/aug/29/academic-publishers-murdoch-socialist.

14. "Publishing, Perishing, and Peer Review," *The Economist*, January 22, 1998, https://www.economist.com/science-and-technology/1998/01/22/publishing-perishing-and-peer-review.

15. Vincent Larivière, Stefanie Haustein, and Philippe Mongeon, "The Oligopoly of Academic Publishers in the Digital Era," *PLoS ONE*, 10, no. 6 (June 10, 2015), https://journals.plos.org/plosone/article?id=10.1371/journal.pone.0127502#pone.0127502.ref047.

Chapter 7: Peer Review and "the Natural Selection of Bad Science"

1. Rex Features, "Sir John Maddox: The Nature of *Nature*," *The Economist*, April 23, 2009, https://www.economist.com/science-and-technology/2009/04/23/the-nature-of-nature; see also retiring editor John Maddox, "Valediction from an Old Hand," *Nature* 378, no. 6557 (1995): 521–23, https://www.nature.com/articles/378521a0.

2. "History of Nature," *Nature*, https://www.nature.com/nature/about/history-of-nature.

3. For a critique of peer review in general, see Bruce Charlton, "The Cancer of Bureaucracy: How It Will Destroy Science, Medicine, Education; and Eventually Everything Else," *Medical Hypotheses* 74, no. 6 (2010): 961–65.

4. Melinda Baldwin, "Scientific Autonomy, Public Accountability, and the Rise of 'Peer Review' in the Cold War United States," *Isis* 109, no. 3 (September 2018), https://www.journals.uchicago.edu/doi/pdf/10.1086/700070.
5. Wikipedia, s.v. "Scholarly Peer Review," https://en.wikipedia.org/wiki/Scholarly_peer_review.
6. Maddox, "Valediction from an Old Hand."
7. Wikipedia, s.v. "Positional Good," https://en.wikipedia.org/wiki/Positional_good.
8. For details, see Wikipedia, s.v. "Impact Factor," https://en.wikipedia.org/wiki/Impact_factor.
9. Daniel Sarewitz, "Saving Science," *New Atlantis* (Spring/Summer 2016).
10. A recent and very thoughtful attempt to improve the validity of citation counts was successfully peer reviewed and published in *Proceedings of the National Academy of Sciences*—only to be withdrawn, not because of any scientific errors but because of political objections by "diversity" fanatics. See Lawrence M. Krauss, "An Astronomer Cancels His Own Research—Because the Results Weren't Popular," Quillette, November 10, 2021, https://quillette.com/2021/11/10/an-astronomer-cancels-his-own-research-because-the-results-werent-popular/.
11. For an exposé of these "publishers" see Wikipedia, s.v. "Bentham Science Publishers," https://en.wikipedia.org/wiki/Bentham_Science_Publishers#Publishing_divisions and links therein.
12. For example, Charles Gross, "Disgrace: On Mark Hauser," *The Nation*, December 21, 2011, https://www.thenation.com/article/archive/disgrace-marc-hauser/; Nadi Bey and Leah Boyd, "Researchers Raise Concerns of Fraud and Ambiguity in Two Studies Authored by Dan Ariely, Renowned Duke Researcher and Professor," *The Chronicle*, August 19, 2021, https://www.dukechronicle.com/article/2021/08/duke-university-dan-ariely-fraudulent-data-colada-research-2012-2004-economics-psychology-statistics.
13. Les Hatton and Gregory Warr, "Scientific Peer Review: An Ineffective and Unworthy Institution," *Times Higher Education*, December 9, 2017, https://www.timeshighereducation.com/blog/scientific-peer-review-ineffective-and-unworthy-institution. See also Stuart Ritchie, *Science Fictions: How Fraud, Bias, Negligence, and Hype Undermine the Search for Truth* (New York: Metropolitan Books, 2020).
14. Hatton and Warr, "Scientific Peer Review: An Ineffective and Unworthy Institution."

15. Paul E. Smaldino and Richard McElreath, "The Natural Selection of Bad Science," *Royal Society Open Science* 3, no. 9 (September 2016), https://royalsocietypublishing.org/doi/full/10.1098/rsos.160384.

16. Tom Bartlett, "Spoiled Science," *Chronicle of Higher Education*, March 17, 2017 https://www.chronicle.com/article/spoiled-science/.

17. Aaron E. Carroll, "The Cookie Crumbles: A Retracted Study Points to a Larger Truth," *New York Times*, October 23, 2017, https://www.nytimes.com/2017/10/23/upshot/the-cookie-crumbles-a-retracted-study-points-to-a-larger-truth.html.

18. Bartlett, "Spoiled Science."

19. Alan Grafen, "William Donald Hamilton: 1 August 1936–7 March 2000," *Biographical Memoirs of Fellows of the Royal Society London* 50 (2004): 109–32, http://users.ox.ac.uk/~grafen/cv/WDH_memoir.pdf.

20. John Staddon, "Reinforcement As Input: Cyclic Interval-Variable Schedule," *Science* 145, no. 3630 (July 1964): 410–12, https://pubmed.ncbi.nlm.nih.gov/14172609/.

21. Randy Schekman, "How Journals like *Nature*, *Cell* and *Science* are Damaging Science," *The Guardian*, December 9, 2013, https://www.theguardian.com/commentisfree/2013/dec/09/how-journals-nature-science-cell-damage-science; Björn Brembs, Katherine Button, and Marcus Munafò, "Deep Impact: Unintended Consequences of Journal Rank," *Frontiers in Human Neuroscience* 7, no. 291 (June 2013), https://www.frontiersin.org/articles/10.3389/fnhum.2013.00291/full.

22. Brembs et al., "Deep Impact."

23. Schekman, "How Journals like *Nature*, *Cell* and *Science*."

24. Ian Sample, "Research Findings That Are Probably Wrong Cited Far More Than Robust Ones, Study Finds," *The Guardian*, May 21, 2021, https://www.theguardian.com/science/2021/may/21/research-findings-that-are-probably-wrong-cited-far-more-than-robust-ones-study-finds.

25. Paul E. Smaldino and Richard McElreath, "The Natural Selection of Bad Science," *Royal Society Open Science* 3, no. 160384 (September 2016), https://doi.org/10.1098/rsos.160384.

26. Richard Harris, *Rigor Mortis: How Sloppy Science Creates Worthless Cures, Crushes Hope, and Wastes Billions* (New York: Basic Books, 2017).

27. Wikipedia, s.v. "Alexandra Elbakyan," https://en.wikipedia.org/wiki/Alexandra_Elbakyan.

28. Sci-Hub, http://ww16.sci-hub.tw/?sub1=20201229-0650-0001-bc0c-1cbdbabf7766.

29. See, for example, Researchers.One, https://researchers.one/. An interesting solution is described by Adam Ellwanger, "An Innovative Solution to the Failures of Peer Review," James G. Martin Center for Academic Renewal, August 13, 2021, https://www.jamesgmartin.center/2021/08/an-innovative-solution-to-the-failures-of-peer-review/.

Chapter 8: Is the Climate Warming? Is There More Extreme Weather?

1. "New Book—Thermal Physics of the Atmosphere," Royal Meteorological Society, November 16, 2020, https://www.rmets.org/news/new-book-thermal-physics-atmosphere.

2. John Staddon and Peter Morcombe, "The Case for Carbon Dioxide," *Academic Questions* 33, no. 2 (May 2020): 246–58.

3. Respected theoretical physicist Freeman Dyson was also suspicious of global warming because of the heat generated, not by CO_2, but by the intolerance of its advocates. John Staddon, "Variation and Diversity: A Tribute to Freeman Dyson," *Academic Questions* 33, no.3 (July 2020): 436–47.

4. See Richard Feynman's wonderful little book: *QED: The Strange Theory of Light and Matter* (Princeton: Princeton University Press, 1985); Wikipedia, s.v. "QED: The Strange Theory of Light and Matter," 2022, https://en.wikipedia.org/wiki/QED:_The_Strange_Theory_of_Light_and_Matter.

5. Wikipedia, s.v. "Global Warming Hiatus," https://en.wikipedia.org/wiki/Global_warming_hiatus.

6. "Global Warming's Boom Town," *The Economist,* May 24, 2007. On the other hand, see Mary Poffenroth, "Fear Appeal in Climate Change Communication: Analysis of The Economist Magazine Covers," *Academia Letters*, Article 634 (May 2021), https://www.academia.edu/49056777/Fear_Appeal_in_Climate_Change_Communication_Analysis_of_The_Economist_Magazine_Covers?email_work_card=thumbnail.

7. Andrew Shepherd, Helen Amanda Fricker, and Sinead Louise Farrell, "Trends and Connections across the Antarctic Cryosphere," *Nature* 558, no. 7709 (June 2018): 223–32; Thomas Slater et al., "Review Article: Earth's Ice Imbalance," *Cryosphere* 15 (2021): 233–46.

8. Bob Berwyn, "Why Is Antarctica's Sea Ice Growing While the Arctic Melts? Scientists Have an Answer," Inside Climate News, May 31, 2016. https://insideclimatenews.org/news/31052016/why-antarctica-sea-ice-level-growing-while-arctic-glaciers-melts-climate-change-global-warming/.

9. Harry Cockburn, "Greenland's Biggest Glacier Suddenly Slows Down and Thickens, Baffling Scientists," *The Independent*, May 15, 2019; Arley Titzler, "Video of the Week: Flying over Jakobshavn Glacier," GlacierHub, May 15, 2019, https://glacierhub.org/tag/ilulissat/.

10. "Peruvian Glaciers Have Shrunk by 30 Percent since 2000," YaleEnvironment360, October 7, 2019, https://e360.yale.edu/digest/peruvian-glaciers-have-shrunk-by-30-percent-since-2000.

11. "Tech Must Help Combat Climate Change, Says Sundar Pichai," *The Economist*, November 17, 2020, https://www.economist.com/the-world-ahead/2020/11/17/tech-must-help-combat-climate-change-says-sundar-pichai.

12. Jennifer A. Dlouhy and Josh Wingrove, "Biden Calls Climate Change 'Existential Threat of Our Time,'" Bloomberg, December 19, 2020, https://www.bloomberg.com/news/articles/2020-12-19/biden-calls-climate-change-existential-threat-of-our-time.

13. President of Harvard University, Lawrence S. Bacow, "Climate Change," *Harvard Magazine*, Fall 2019, https://www.harvardmagazine.com/2019/09/climate-change; and more recently, Lawrence S. Bacow, "A Message from President Bacow on Climate Change," Harvard: Office of the President, April 21, 2020, https://www.harvard.edu/president/news/2020/message-from-president-bacow-on-climate-change.

14. Oliver Milman and David Smith, "'Listen to the Scientists': Greta Thunberg Urges Congress to Take Action, *The Guardian*, September 18, 2019, https://www.theguardian.com/us-news/2019/sep/18/greta-thunberg-testimony-congress-climate-change-action?ref=hvper.com. See also Lauren Gambino, "Greta Thunberg to Congress: 'You're Not Trying Hard Enough. Sorry,'" *The Guardian*, September 17, 2019, https://www.theguardian.com/environment/2019/sep/17/greta-thunberg-to-congress-youre-not-trying-hard-enough-sorry.

15. Edwin X. Berry, "Preprint #2: The Physics Model Carbon Cycle for Human CO_2," EdBerry.com, November 25, 2021, https://edberry.com/blog/climate/climate-physics/human-co2-has-little-effect-on-the-carbon-cycle/.

16. Watts Up with That?, by Anthony Watts, 2006–2021, https://wattsupwiththat.com/.

17. Skeptical Science, by John Cook, 2022, https://skepticalscience.com/.

18. Joseph Bast and Roy Spencer, "The Myth of the Climate Change '97%,'" *Wall Street Journal*, May 26, 2014, https://www.wsj.com/articles/joseph-bast-and-roy-spencer-the-myth-of-the-climate-change-97-1401145980; see also

"Global Warming Petition Project," PetitionProject.org, http://www.
petitionproject.org/index.php.

19. John Staddon, "Diet Reporting—the Real Fake News," Quillette, September
18, 2019, https://quillette.com/2019/09/18/diet-reporting-the-real-fake-news/.

20. Jeff Tollefson, "The Hard Truths of Climate Change—by the Numbers,
Nature 573 (September 2019): 324–27, https://www.nature.com/immersive/
d41586-019-02711-4/index.html.

21. It is literally impossible to connect any *individual* extreme weather event to
the secular global temperature change implied by the anthropogenic global
warming hypothesis. This a problem for litigators, and it has led to the
invention of a new label—*attribution science*—which suggests that new tools
now exist to connect climate to weather: "Developments in attribution
science are improving our ability to detect human influence on extreme
weather events. By implication, the legal duties of government, business and
others to manage foreseeable harms are broadening, and may lead to more
climate change litigation." Sophie Marjanac, Lindene Patton, and James
Thornton, "Acts of God, Human Influence and Litigation," *Nature Geoscience*
10 (2017): 616–19. The new label may help persuade a skeptical court; it adds
nothing scientifically.

22. No, says physicist Steven Koonin in *Unsettled: What Climate Science Tells Us,
What It Doesn't, and Why It Matters* (Dallas: BenBella Books, 2021). See also
Mark P. Mills, "'Unsettled' Review: The 'Consensus' on Climate," *Wall Street
Journal*, April 25, 2021, https://www.wsj.com/articles/unsettled-review-
theconsensus-on-climate-11619383653.

23. "Cycles of Atlantic Hurricanes; Global Warming and Hurricanes; Damage
and Death Trends," http://www.atmo.arizona.edu/students/courselinks/
fall16/atmo336/lectures/sec2/hurricanes3.html.

24. Philip J. Klotzbach et al., "Continental U.S. Hurricane Landfall Frequency
and Associated Damage: Observations and Future Risks," *Bulletin of the
American Meteorological Society* 99, no. 7 (July 2018): 1359–76, https://
journals.ametsoc.org/view/journals/bams/99/7/bams-d-17-0184.1.xml.

25. Paul Homewood, "Explaining the Extreme Weather Events That Did Not
Happen," Not a Lot of People Know That, March 17, 2016, https://
notalotofpeopleknowthat.wordpress.com/2016/03/17/explaining-the-
extreme-weather-events-that-did-not-happen/

26. National Oceanic and Atmospheric Administration, "Temperature Change
and Carbon Dioxide Change," https://www.ncei.noaa.gov/sites/default/

files/2021-11/8%20-%20Temperature%20Change%20and%20Carbon%20
Dioxide%20Change%20-%20FINAL%20OCT%202021.pdf.

27. Michael E. Mann, Raymond S. Bradley, and Malcolm K. Hughes, "Northern
    Hemisphere Temperatures During the Past Millennium: Inferences,
    Uncertainties, and Limitations," *Geophysical Research Letters* 26, no. 6 (March
    1999): 759–62, https://agupubs.onlinelibrary.wiley.com/doi/
    pdf/10.1029/1999GL900070. For further discussion of the "hockey-stick"
    graph, see "Temperature Variations in Past Centuries and the So-Called
    'Hockey Stick,'" RealClimate, December 4, 2004, http://www.realclimate.org/
    index.php/archives/2004/12/temperaturevariations-in-past-centuries-and-
    the-so-called-hockey-stick/.

28. The site Just Facts, https://www.justfacts.com/globalwarming.asp#media-
    scientific, has an entry on global warming which describes in some detail
    difficulties in inferring local (never mind global) temperature in the absence
    of modern measurement methods (i.e., more than 150 years or so ago). The
    site also has numerous quotes from the contentious history of the GW issue
    as well as much historical climate data.

29. Wikipedia, s.v. "Hockey Stick Graph," https://en.wikipedia.org/wiki/Hockey_
    stick_graph.

30. Bradely A. Benbrook et al., "Competitive Enterprise Institute and *National
    Review* v. Michael E. Mann," Cato Institute, July 8, 2019, https://www.cato.
    org/publications/legal-briefs/competitive-enterprise-institute-national-
    review-v-michael-e-mann.

31. Wikipedia, "North Report, last modified October 18, 2021, https://
    en.wikipedia.org/wiki/North_Report.

32. Ross McKitrick, "A Brief Retrospective on the Hockey Stick," for inclusion in
    the compendium volume *Climate Change: The Facts 2014* (Australia: Institute
    for Policy Analysis), May 23, 2014, https://www.rossmckitrick.com/
    uploads/4/8/0/8/4808045/hockey-stick-retrospective.pdf.

33. *Encyclopedia Britannica Online*, s.v. "Medieval Warm Period," by John P.
    Rafferty, November 18, 2014, https://www.britannica.com/science/medieval-
    warm-period.

34. See also Wikipedia, s.v. "Geologic Temperature Record," https://en.wikipedia.
    org/wiki/Geologic_temperature_record.

35. For more details on the data behind the graph, see Dieter Lüthi et al., "High-
    Resolution Carbon Dioxide Concentration Record 650,000–800,000 Years
    before Present," *Nature* 453 (2008): 379–82, https://www.nature.com/articles/
    nature06949; Jean Jouzel et al., "Orbital and Millenial Antarctic Climate

Variability over the Past 800,000 Years," *Science* 317, no. 5839: 793–96, https://
www.researchgate.net/publication/6223217_Orbital_and_Millennial_
Antarctic_Climate_Variability_over_the_Past_800000_Years.

Chapter 9: Carbon Dioxide and the AGW Hypothesis

1. National Oceanic and Atmospheric Administration, "Temperature Change and Carbon Dioxide Change," October 2021, https://www.ncdc.noaa.gov/global-warming/temperature-change.
2. Kathryn Hansen, "Water Vapor Confirmed as Major Player in Climate Change," NASA, November 17, 2008, https://www.nasa.gov/topics/earth/features/vapor_warming.html.
3. Wikipedia, s.v. "Keeling Curve," https://en.wikipedia.org/wiki/Keeling_Curve#/media/File:Mauna_Loa_CO2_monthly_mean_concentration.svg.
4. "Monthly Mean Carbon Dioxide Measured at Mauna Loa Observatory, Hawaii," JunkScience, September 6, 2012, http://junksciencearchive.com/MSU_Temps/MaunaLoaCO2.png.
5. "How Do Human CO_2 Emissions Compare to Natural CO_2 Emissions?," Skeptical Science, updated July 2015, https://skepticalscience.com/human-co2-smaller-than-natural-emissions.htm.
6. "Water Vapor and Climate Change: ACS Climate Science Toolkit: Narratives," ACS Chemistry for Life, https://www.acs.org/content/acs/en/climatescience/climatesciencenarratives/its-water-vapor-not-the-co2.html.
7. This is not as implausible as it sounds: after all, a few drops of India ink can turn a glass of clear water dark if not completely black. Perhaps CO_2 has the same effect in the atmosphere, albeit with infrared rather than visible wavelengths. But as a NASA site comments: "The exact amount of the energy imbalance [darkening] is very hard to measure . . ." "Climate Forcings and Global Warming," NASA Earth Observatory, January 14, 2009, https://earthobservatory.nasa.gov/features/EnergyBalance/page7.php.
8. Wikipedia, s.v. "Greenhouse Gas," https://en.wikipedia.org/wiki/Greenhouse_gas.
9. Robert McSweeney and Zeke Hausfather, "Q&A: How Do Climate Models Work?," CarbonBrief, January 15, 2018, https://www.carbonbrief.org/qa-how-do-climate-models-work; Wikipedia, s.v. "Climate Model," https://en.wikipedia.org/wiki/Climate_model.
10. Described here: Larry M., "How Reliable Are Climate Models?," Skeptical Science, https://skepticalscience.com/climate-models-intermediate.htm; but one skeptic points to the increasingly divergent predictions as models

continue to evolve. Holman W. Jenkins Jr., "How a Physicist Became a Climate Truth Teller," *Wall Street Journal*, April 16, 2021, https://www.wsj.com/articles/how-a-physicist-became-a-climate-truth-teller-11618597216.

11. J. David Neelin et al., "Considerations for Parameter Optimization and Sensitivity in Climate Models," *Proceedings of the National Academy of Sciences of the United States of America* 107, no. 50 (December 2010): 21349–54, https://www.pnas.org/content/pnas/early/2010/11/22/1015473107.full.pdf.

12. Ross McKitrick, "The IPCC's Attribution Methodology Is Fundamentally Flawed," Climate Etc., August 18, 2021, https://judithcurry.com/2021/08/18/the-ipccs-attribution-methodology-is-fundamentally-flawed/; and see Steven Koonin, *Unsettled: What Climate Science Tells Us, What It Doesn't, and Why It Matters* (Dallas: BenBella Books, 2021) in Mark P. Mills, "'Unsettled' Review: The 'Consensus' on Climate," *Wall Street Journal*, April 25, 2021, https://www.wsj.com/articles/unsettled-review-theconsensus-on-climate-11619383653.

13. John Droz, "Climate Change Computer Modes: The Good, the Bad, & the Ugly," revised August 1, 2020, http://wiseenergy.org/Energy/AGW/Computer_Models_Abbreviated.pdf.

14. "Hurricane Forecast Model Accuracy," Hurricanes: Science and Society, http://www.hurricanescience.org/science/forecast/models/modelskill/.

15. Michael Lemonick, "Freeman Dyson Takes on the Climate Establishment," Yale Environment 360, June 4, 2009, https://e360.yale.edu/features/freeman_dyson_takes_on_the_climate_establishment. Syukuro Manabe shared the 2021 Nobel Prize with two other climate physicists.

16. Wikipedia, s.v. "Three-Body Problem," https://en.wikipedia.org/wiki/Three-body_problem.

17. Current models are diverging in their predictions: Jenkins Jr., "How a Physicist Became a Climate Truth Teller."

18. Wikipedia, s.v. "Black-Scholes Model," https://en.wikipedia.org/wiki/Black%E2%80%93Scholes_model.

19. "The Midas Formula," BBC, December 2, 1999, http://www.bbc.co.uk/science/horizon/1999/midas.shtml.

20. Svante Arrhenius, "The Probable Cause of Climate Fluctuations," Friends of Science Society, 2014, https://www.friendsofscience.org/assets/documents/Arrhenius%201906,%20final.pdf.

21. Elisabeth Crawford, "Arrhenius' 1896 Model of the Greenhouse Effect in Context," *Ambio* 26, no. 1, (February 1997): 6–11.

22. Variation in solar output is an obvious possibility for non-AGW changes in the temperature of the earth. See, for example, Nicola Scafetta, "Solar and Planetary Oscillation Control on Climate Change: Hind-Cast, Forecast and a Comparison with the CMIP5 GCMS," *Energy & Environment* 24, no. 3–4 (June 2013); and Ronan Connolly et al., "How Much Has the Sun Influenced Northern Hemisphere Temperature Trends? An Ongoing Debate," *Research in Astronomy and Astrophysics* 21, no. 6. (2021).

23. "CO_2 Lags Temperature—What Does It Mean?," Skeptical Science, April 21, 2021, https://skepticalscience.com/co2-lags-temperature.htm. Skeptical Science, a site critical of AGW skeptics, discusses various interpretations of this lag.

24. Abel Barral et al., "CO_2 and Temperature Decoupling at the Million-Year Scale during the Cretaceous Greenhouse," *Scientific Reports* 7, no. 8310 (August 2017), https://www.nature.com/articles/s41598-017-08234-0.

25. Figure 9 is found in "Hemlata Pant et al., "Innovations in Agriculture, Environment and Health Research for Ecological Restoration" (conference paper for Proceedings of National Symposium, 2019), https://www.researchgate.net/publication/342589030_Innovations_in_Agriculture_Environment_and_Health_Research_for_Ecological_Restoration.

26. Keith Johnson, "How Carbon Dioxide Became a 'Pollutant,'" *Wall Street Journal*, April 18, 2009, https://www.wsj.com/articles/SB124001537515830975.

27. Wikipedia, s.v. "Free-Air Concentration Enrichment," https://en.wikipedia.org/wiki/Free-air_concentration_enrichment.

28. Samson Reiny, "CO_2 Is Making Earth Greener—for Now," NASA's Earth Science News Team, April 26, 2016, https://climate.nasa.gov/news/2436/co2-is-making-earth-greenerfor-now/; Samson Reiny, "Carbon Dioxide Fertilization Greening Earth, Study Finds," NASA's Earth Science News Team, April 26, 2016, https://www.nasa.gov/feature/goddard/2016/carbon-dioxide-fertilization-greening-earth.

29. Reiny, "Carbon Dioxide Fertilization."

30. Doyle Rice, "Study: Cold Kills 20 Times More People than Heat," *USA Today*, updated May 21, 2015, https://www.usatoday.com/story/weather/2015/05/20/cold-weather-deaths/27657269/; Alex Berezow, "More People Die in Winter than Summer, American Council on Science and Health, July 10, 2019, https://www.acsh.org/news/2019/07/10/more-people-die-winter-summer-14146#:~:text=Specifically%2C%20people%20are%20far%20likelier,media%20hypes%20summer%20heat%20waves.

31. Christopher Booker: *Global Warming: A Case Study in Groupthink; How Science Can Shed New Light on the Most Important 'Non-Debate' of Our Time* (London: The Global Warming Policy Foundation, 2018), https://www. thegwpf.org/content/uploads/2018/02/Groupthink.pdf.

Chapter 10: Killing the Messenger

1. Alvin Weinberg "Science and Trans-Science," *Minerva* 10, no. 2 (1972): 209–22.
2. Published as Peter Arcidiacono, Esteban M. Aucejo, and Ken Spenner, "What Happens After Enrollment? An Analysis of the Time Path of Racial Differences in GPA and Major Choice," *IZA Journal of Labor Economics* 1, no. 5 (October 2012).
3. Peter Schmidt, "Study Disputes Claims That Preferentially Admitted Students Catch Up," *Chronicle of Higher Education,* January 10, 2012.
4. Students admitted not on academic merit alone, but as children of alumni and others with a connection to the university.
5. This is another example of language policing: "weaker backgrounds" is now regularly used as a euphemism for "weaker students." We know the students are "weaker" by their grade point average. That's the definition of "weak." Maybe their poor performance is caused entirely by their academic background, or maybe it has endogenous causes; they may be lazy, distracted, or dumb, for example. "Weak students" would be a more neutral way to refer to these kids than saying they have a "weak academic background," which assumes that we know something that in fact we do not, namely the cause of their low GPA. This kind of language taboo is now endemic in social science and loads the dice in favor of exogenous causes for disparities.
6. Mike Shammas, "Unpublished Study Draws Ire from Minorities," *Duke Chronicle*, January 17, 2012, https://www.dukechronicle.com/article/ unpublished-study-draws-ire-minorities.
7. Richard Brodhead, "Brodhead: Duke and the Legacy of Race," *Duke Today*, March 22, 2012, https://today.duke.edu/2012/03/rhbfacultytalk.
8. Here is another example that happens to be again from Duke, but Duke is neither alone nor even the worst case: John Staddon, "Duke Divinity School's Race to the Bottom," James G. Martin Center for Academic Renewal, April 16, 2018, https://www.jamesgmartin.center/2018/04/duke-divinity-schools-race-to-the-bottom/.

Chapter 11: The Devolution of Social Science: How the Fragmentation of Sociology Has Led to Absurdity

1. Examples from a Royal Society publication, *Miscellanea Curiosa*, of 1708 also contains several mathematical contributions from Halley: *Miscellanea Curiosa*, 2nd ed. (London: 1708), https://www.biodiversitylibrary.org/creator/126992#/titles.

2. Oliver Lodge, *Past Years: An Autobiography* (London: Hodder and Stoughton, 1931), 135.

3. "Scientific Sections," British Science Association, https://www.britishscienceassociation.org/scientific-sections.

4. Peder Olesen Larsen and Markus von Ins, "The Rate of Growth in Scientific Publication and the Decline in Coverage Provided by Science Citation Index," *Scientometrics* 84, no. 3 (2010): 575–603, https://www.ncbi.nlm.nih.gov/pmc/articles/PMC2909426/.

5. Émile Durkheim, *The Rules of Sociological Method: And Seleted Texts on Sociology and Its Method* (Paris: 1895; London: Macmillan, 1982), 33.

6. Ibid., 50.

7. Ibid., 50–51.

8. Ibid., 52.

9. Eduardo Bonilla-Silva, *Racism without Racists: Color-Blind Racism and the Persistence of Racial Inequality in the Unites States*, 4th ed. (Lanham, Maryland: Rowman and Littlefield, 2013), 21.

10. Tukufu Zuberi, "Deracializing Social Statistics: Problems in the Quantification of Race," *The ANNALS of the American Academy of Political and Social Science* 568, no. 1 (March 2000): 177.

11. Wikipedia, s.v. "Four Causes," https://en.wikipedia.org/wiki/Four_causes.

12. For the application of these distinctions in behavioral science, see John Staddon, *The New Behaviorism: Foundations of Behavioral Science*, 3rd ed. (Philadelphia: Psychology Press, 2021).

13. Wikipedia, s.v. "Eugenics," https://en.wikipedia.org/wiki/Eugenics.

14. Zuberi, "Deracializing Social Statistics," 174.

15. Erika Check Hayden, "Ethics: Taboo Genetics," *Nature* 502 (October 2013): 26–28. https://www.nature.com/news/ethics-taboo-genetics-1.13858. See also Mark Mercer, "Why Scholars Won't Research Group Differences," *Academic Questions* 33 (Winter 2020): 581–85.

16. Monnica T. Williams, "Colorblind Ideology Is a Form of Racism," *Psychology Today*, December 27, 2011, https://www.psychologytoday.com/us/blog/culturally-speaking/201112/colorblind-ideology-is-form-racism.

Chapter 12: Contemporary Sociology: Race and Ethnic Studies

1. Wikipedia, s.v. "The Protestant Ethic and the Spirit of Capitalism," https://en.wikipedia.org/wiki/The_Protestant_Ethic_and_the_Spirit_of_Capitalism

2. Alan D. Sokal, "Transgressing the Boundaries: Towards a Transformative Hermeneutics of Quantum Gravity," *Social Text* 46/57 (Spring/Summer 1996): 217–52. See also Wikipedia, s.v. "Sokal Affair, https://en.wikipedia.org/wiki/Sokal_affair; for more on postmodernist and science-studies hoaxes, see Alan Sokal, "What the 'Conceptual Penis' Hoax Does and Does Not Prove," *Chronicle of Higher Education*, June 14, 2017, https://www.chronicle.com/article/what-the-conceptual-penis-hoax-does-and-does-not-prove/; and, more recently, Tom Bartlett, "The Mysterious Case of the Nonsense Papers: A Peer-Reviewed Journal Published Hundreds of Them. Why?," *Chronicle of Higher Education* , September 28, 2021, https://www.chronicle.com/article/why-did-a-peer-reviewed-journal-publish-hundreds-of-nonsense-papers.

3. Tukufu Zuberi and Eduardo Bonilla-Silva, "Towards a Definition of White Logic and White Methods," in *White Logic, White Methods: Racism and Methodology*, ed. Tukufu Zuberi and Eduardo Bonilla-Silva (Lanham, Maryland: Rowman and Littlefield, 2008), 17.

4. Eduardo Bonilla-Silva, "What We Were, What We Are, and What We Should Be: The Racial Problem of American Sociology," *Social Problems* 64, no. 2 (May 2017): 183. Perhaps this dismissal of objectivity is just a rhetorical flourish that one need not take very seriously. Alas, no; it has become a cause in itself, even to the point that mathematics, by definition the most objective science, is being unraveled. For example, the Oregon Department of Education in February 2021, commemorating Black History Month, sent out a newsletter promoting "A Pathway to Math Equity Micro-Course," EquitableMath.org, https://equitablemath.org/. Five downloads, lavishly illustrated with pictures of children of color, are offered, ranging from "Dismantling Racism on Mathematics" to "Sustaining Equitable Practice." "While primarily for math educators, this text advocates for a collective approach to dismantling white supremacy." Amid thousands of words of what sounds more like psychotherapy/propaganda than math instruction— "Engage in self-examination of teaching mathematics with an equity lens," and "guidance and resources for educators to use now as they plan their curriculum, while also offering opportunities for ongoing self-reflection as they seek to develop an anti-racist math practice"—it is tough to find any references to actual mathematics or solving problems. The best I could find was this: "White supremacy culture shows up in math classrooms when . . .

the focus is only on getting the right answer." Correct math is racist, apparently, or at least "white." One wonders who is funding this costly drivel.

5. Zuberi and Bonilla-Silva, "Towards a Definition of White Logic and White Methods," in *White Logic, White Methods*, 23.

6. Roger Kiska, "Antonio Gramsci's Long March through History," Acton Institute, December 12, 2019, https://www.acton.org/religion-liberty/volume-29-number-3/antonio-gramscis-long-march-through-history.

7. Zuberi and Bonilla-Silva, "Telling the Real Tale of the Hunt," in *White Logic, White Methods*, 332.

8. Ibid.

9. Wikipedia, s.v. "Mandy Rice-Davies," https://en.wikipedia.org/wiki/Mandy_Rice-Davies.

10. Anna-Esther Younes, review of *White Logic, White Methods: Racism and Methodology*, eds. Tukufu Zuberi and Eduardo Bonilla-Silva, *Graduate Journal of Social Science* 9, no. 1 (March 2012): 92, http://www.gjss.org/sites/default/files/issues/chapters/bookreviews/Journal-09-01--05-Younes.pdf. ("Tukufu" is misspelled "Tufuku" by Younes.)

11. Zuberi and Bonilla-Silva, "Towards a Definition of White Logic and White Methods," in *White Logic, White Methods*, 4.

12. Ibid.

13. Eduardo Bonilla-Silva, *Racism without Racists*, 4th ed. (Lanham, Maryland: Rowman and Littlefield, 2013), ix.

14. Bonilla-Silva, "What We Were."

15. Zuberi and Bonilla-Silva, "Towards a Definition of White Logic and White Methods," in *White Logic, White Methods*, 5.

16. Ibid.

17. Franics Galton, *Hereditary Genius: An Inquiry into Its Laws and Consequences*, 2nd ed. (London: MacMillan, 1914), x, 338, 342.

18. Galton, *Hereditary Genius*, x.

19. For contrasting views, see Sarah A. Tishkoff and Kenneth K. Kidd, "Implications of Biogeography of Human Populations for 'Race' and Medicine," *Nature Genetics* 36, no. 11 (November 2004): S21–S27; and Nicholas Wade, *A Troublesome Inheritance: Genes, Race and Human History*, (New York: Penguin, 2014).

20. Richard Lynn and Gerhard Meisenberg, "The Average IQ of Sub-Saharan Africans: Comments on Wicherts, Dolan, and van der Maas," *Intelligence* 38, no. 1 (January–February 2010): 21–29, https://www.sciencedirect.com/science/article/abs/pii/S0160289609001275.

21. Zuberi and Bonilla-Silva, "Towards a Definition of White Logic and White Methods," in *White Logic, White Methods*, 8.

22. Nicholas Gillham, "Cousins: Charles Darwin, Sir Francis Galton and the Birth of Eugenics," *Significance* (September 2009): 132–35, https://rss. onlinelibrary.wiley.com/doi/epdf/10.1111/j.1740-9713.2009.00379.x

23. Eduardo Bonilla-Silva, *Racism without Racists: Color-Blind Racism and the Persistence of Racial Inequality In the United States*, 2nd ed. (Lanham, Maryland: Rowman & Littlefield, 2006), 212.

24. Ashley "Woody" Doane and Eduardo Bonilla-Silva, eds., *White Out: The Continuing Significance of Racism* (New York: Routledge, 2003), 13.

25. Bonilla-Silva, *Racism without Racists*, 131–50, 264.

26. A nearly identical form of this quotation occurs on this website: "Brief Bio," Professor George Yancy, Ph.D., http://georgeyancy.com/bio.html.

27. Thomas Hobbes, *Leviathan* (New York: Oxford University Press, 1998), 31.

28. Michael Tye, "Qualia," *The Stanford Encyclopedia of Philosophy*, Edward N. Zalta, ed., Fall 2021, https://plato.stanford.edu/entries/qualia/.

29. See, for example, John Staddon, *The New Behaviorism: Foundations of Behavioral Science*, 3rd ed. (Philadelphia: Psychology Press, 2021), especially discussion of the color-phi problem.

30. John Ziman, *Public Knowledge: The Social Dimension of Science* (Cambridge: Cambridge University Press, 1969).

31. Staddon, *The New Behaviorism*.

32. Bonilla-Silva, *Racism without Racists*, 7.

33. See appendix and Jesse Singal, "Psychology's Favorite Tool for Measuring Racism Isn't Up to the Job," *New York* January 11, 2017, https://www.thecut. com/2017/01/psychologys-racism-measuring-tool-isnt-up-to-the-job.html.

34. Bonilla-Silva, *Racism without Racists*.

35. Joshua Rothman, "The Origins of 'Privilege,'" *New Yorker*, May 12, 2014, https://www.newyorker.com/books/page-turner/the-origins-of-privilege.

36. Peggy McIntosh, "White Privilege and Male Privilege: A Personal Account of Coming to See Correspondences through Work in Women's Studies," Wellesley Centers for Women (1988), https://www.wcwonline.org/images/ pdf/White_Privilege_and_Male_Privilege_Personal_Account-Peggy_ McIntosh.pdf; see also Peggy McIntosh, "White Privilege: Unpacking the Invisible Knapsack," *Peace and Freedom* (July/August 1989).

37. For details on Peggy McIntosh's actual privilege see William Ray, "Unpacking Peggy McIntosh's Knapsack," Quillette, August 29, 2018, https://quillette. com/2018/08/29/unpacking-peggy-mcintoshs-knapsack/.

38. "About Me," Robin DiAngelo, PhD—Critical Racial & Social Justice Education, https://www.robindiangelo.com/about-me/.
39. Robin J. DeAngelo, *White Fragility: Why It's So Hard for White People to Talk about Racism* (Boston: Beacon Press, 2018); Michael Eric Dyson praising *White Fragility* at https://www.amazon.com/White-Fragility-People-About-Racism/dp/0807047406/ref=tmm_hrd_swatch_0.
40. Daniel Bergner, "'White Fragility' Is Everywhere. But Does Antiracism Training Work?," *New York Times*, July 15, 2020, https://www.nytimes.com/2020/07/15/magazine/white-fragility-robin-diangelo.html.
41. Claudia Rankine praising DiAngelo's *White Fragility* at https://www.amazon.com/White-Fragility-People-About-Racism/dp/0807047406/ref=tmm_hrd_swatch_0.
42. Katy Waldman, "A Sociologist Examines the 'White Fragility' That Prevents White Americans from Confronting Racism," *New Yorker*, July 23, 2018, https://www.newyorker.com/books/page-turner/a-sociologist-examines-the-white-fragility-that-prevents-white-americans-from-confronting-racism.
43. Robin DiAngelo, "White Fragility," *International Journal of Critical Pedagogy* 3, no. 3 (2011): 54. https://libjournal.uncg.edu/ijcp/article/viewFile/249/116.
44. Ibid., 54–55.
45. Any challenge like this is likely to elicit a hostile reaction: Adam Stanaland and Sarah Gaither, "'Be a Man': The Role of Social Pressure in Eliciting Men's Aggressive Cognition," *Personality and Social Psychology Bulletin* 47, no. 11 (January 2021): 1596–611, https://doi.org/10.1177/0146167220984298.
46. DiAngelo, "White Fragility," 55, 62.
47. See, for example, John Staddon, *The New Behaviorism: Foundations of Behavioral Science*, 3rd ed. (Philadelphia: Psychology Press, 2021).
48. DiAngelo, "White Fragility," 56.
49. Ibid.
50. Ibid.
51. Thomas Sowell, *Discrimination and Disparities*, rev. ed. (New York: Basic Books, 2019). See also Gene Dattel, "Looking Beyond the Numbers," *Academic Questions* 31, no. 3 (September 2018): 368–72.
52. *Why It's So Hard for White People to Talk About Racism* is the subtitle of DiAngelo's book. The obvious answer, which goes unmentioned, is that they are afraid of being called "racist"—which Ibram Kendi, at least, would consider clear proof that they actually are.
53. Catherine MacKinnon, quoted in Daniel A. Farber and Suzanna Sherry, *Beyond All Reason: The Radical Assault on Truth in American Law* (Oxford:

Oxford University Press, 1997). Science is science; to call it "male" is almost meaningless, unless MacKinnon just means that most—far from all—scientists are male; in which case, who cares?

54. CRT star Kimberlé Crenshaw summarizes the first ten years of CRT: "The First Decade: Critical Reflections, or 'A Foot in the Closing Door,'" *UCLA Law Review* 49, no. 5 (2002): 1343–72.

55. Janel George, "A Lesson on Critical Race Theory," *Human Rights* 46, no. 2 (January 2011).

56. Richard Delgado and Jean Stefancic *Critical Race Theory: An Introduction*, 2nd ed. (New York: NYU Press, 2012). Oppression is assumed throughout, as in these leading questions for students: "Would it not be logical for blacks, Latinos, Asians, and Native Americans to unite in one powerful coalition to confront the power system that is oppressing them all?" (95) Also assumed is the racism of color blindness: "Would a determined campaign by every white in this country to be color blind—to completely ignore the race of other people—eliminate the scourge of racism and racial subordination? Or is racism so embedded in our social structures, rules, laws, language, and ways of doing things that the system of white-over-black/brown/yellow subordination would continue . . . ?" (138)

57. Richard A. Jones, "Philosophical Methodologies of Critical Race Theory," APA Blog, August 20, 2019, https://blog.apaonline.org/2019/08/20/philosophical-methodologies-of-critical-race-theory/.

58. Kimberlé Crenshaw, quoted by Richard A. Jones in "Philosophical Methodologies of Critical Race Theory."

59. Richard Delgado, "When a Story Is Just a Story: Does Voice Really Matter?," *Virginia Law Review* 76, no. 1 (February 1990): 95.

60. Ibid.

61. Tukufu and Bonilla-Silva, eds., *White Logic, White Methods*.

62. Roger Scruton, "The Plague of Sociology," *The Times*, October 8, 1985, https://www.fortfreedom.org/l16.htm.

Chapter 13: Systemic Racism: What Do Racial Disparities Really Mean?

1. Bret Stephens, "The Outrage over Sarah Jeong," *New York Times*, August 9, 2018, https://www.nytimes.com/2018/08/09/opinion/sarah-jeong-tweets-opinion-section.html. Interestingly, "race traitors," whites who make racist comments about whites seem to be less tolerated: I. K., "Can White People Experience Racism: Accusations of 'Reverse Racism' Haunt an American

Professor," *The Economist*, September 18, 2018, https://www.economist.com/open-future/2018/09/18/can-white-people-experience-racism; and Michael I. Norton and Samuel R. Sommers, "Whites See Racism as a Zero-Sum Game That They Are Now Losing," *Psychological Science* 6, no. 3 (May 2011): 215–18.

2.　The terms *institutional* and *structural* racism are also used.

3.　"What Is Systemic Racism?," United States Conference of Catholic Bishops, 2018, https://www.usccb.org/issues-and-action/human-life-and-dignity/racism/upload/racism-and-systemic-racism.pdf.

4.　"Biden: 'Systemic Racism Is Corrosive, Destructive and Costly,'" BBC, January 27, 2021, https://www.bbc.com/news/av/world-us-canada-55820709.

5.　Matthew J. Franck, "Racism Is Real. But Is 'Systemic Racism'"? Public Discourse, September 14, 2020, https://www.thepublicdiscourse.com/2020/09/71461/.

6.　Nicole Daniels, "What States Are Saying about Race and Racism in America," *New York Times*, March 2, 2021, https://www.nytimes.com/2021/02/18/learning/what-students-are-saying-about-race-and-racism-in-america.html.

7.　Ashley "Woody" Doane, "Shades of Colorblindness: Rethinking Racial Ideology in the United States," in *The Colorblind Screen: Television in Post-Racial America*, ed. Sarah Nilsen and Sarah E. Turner (Oxford: University Press Scholarship Online, 2016), https://www.universitypressscholarship.com/view/10.18574/nyu/9781479809769.001.0001/upso-9781479809769-chapter-1.

8.　Claire Cain Miller et al., "'When I See Racial Disparities, I See Racism.' Discussing Race, Gender and Mobility," *New York Times*, March 27, 2018, https://www.nytimes.com/interactive/2018/03/27/upshot/reader-questions-about-race-gender-and-mobility.html..

9.　Kendi has his own logic, interpreting denial as proof of racism. Jim Mandelaro, "Ibram X. Kendi: 'The Very Heart of Racism Is Denial,'" University of Rochester, February 25, 2021, https://www.rochester.edu/newscenter/ibram-x-kendi-the-very-heartbeat-of-racism-is-denial-470332/; Kendi also claims that there can be "no neutrality in the racism struggle." Ibram X. Kendi, *How to Be an Antiracist* (New York: One World, 2019), 9. (The *New York Times* call this a "stunner of a book.") On the other hand, his tautologous definition of "racist" leaves much room for maneuver: a "racist" is just "one who is supporting a racist policy . . ." (13). Logic surfaces when Kendi acknowledges that the existence of endogenous race differences would sink his scheme: "To be antiracist is to think nothing is behaviorally wrong or right—inferior or superior—with any of the racial groups." (105). Apparently individual differences are *verboten* unless proportionately distributed.

10. Andrew Koppleman, "What Is Systemic Racism Anyway?," *USA Today* , September 23, 2020, https://www.usatoday.com/story/opinion/2020/09/23/systemic-racism-how-really-define-column/5845788002/.

11. "What Is Systemic Racism? [Videos]," Race Forward, https://www.raceforward.org/videos/systemic-racism.

12. For example, *Discrimination and Disparities*, rev. ed. (New York: Basic Books, 2019), *Black Rednecks and White Liberals* (New York: Encounter Books, 2006), and *A Conflict of Visions: Ideological Origins of Political Struggles* rev. ed. (New York: Basic Books, 2007).

13. "Kendi's goals are openly totalitarian," comments Coleman Hughes in "How to Be an Anti-intellectual: A Lauded Book about Anti-Racism Is Wrong on Its Facts and in Its Assumptions," *City Journal*, October 27, 2019, https://www.city-journal.org/how-to-be-an-antiracist. Such goals will not work, of course: Can you name a totalitarian state where all are equal? See also Glenn Loury, "You Have Two Alternatives. You Can Live with Disparities, or You Can Live in Totalitarianism," American Enterprise Institue, July 25, 2020, https://www.aei.org/carpe-diem/you-have-two-alternatives-you-can-live-with-disparities-or-you-can-live-in-totalitarianism/.

14. See, for example, Peter S. Magnusson, "Is IQ a Predictor of Success?," *Forbes*, September 16, 2015, https://www.forbes.com/sites/quora/2015/09/16/is-iq-a-predictor-of-success/?sh=35f88d073604; University of Tennessee at Knoxville, "IQ Is a Better Predictor of Adult Economic Success than Math," Science Daily, March 14, 2019, https://www.sciencedaily.com/releases/2019/03/190314111017.htm, and many other references. Columbia legal scholar Jamal Greene agrees, but thinks that the law should mitigate the "unfairnesses" due to the way society favors people with a capacity for "logical-mathematical and verbal-linguistic intelligence," i.e., IQ. Kalefa Sanneh, "From Guns to Gay Marriage, How Did Rights Take Over Politics?: The N.R.A., the Supreme Court, and the Forces Driving the Country's Most Intractable Legal Debates," *New Yorker*, May 24, 2021, https://www.newyorker.com/magazine/2021/05/31/from-guns-to-gay-marriage-how-did-rights-take-over-politics.

15. Alan S. Kaufman, *IQ Testing 101* (New York: Springer, 2009), especially chapter 7.

16. For a discussion of the limitations of statistical heritability, see John Staddon, *Scientific Method: How Science Works, Fails to Work or Pretends to Work*,

(London: Taylor and Francis, 2017), chapter 4; Wikikpedia, s.v. "Heritability of IQ, https://en.wikipedia.org/wiki/Heritability_of_IQ.

17. Controversially pointed out by Richard J. Herrnstein in "IQ," *Atlantic Monthly* (September 1971): 43–64, although he was far from the first; Francis Galton and Lyman Tower Sargent, "The Eugenic College of Kantsaywhere," *Utopian Studies* 12, no. 2 (2001): 191–209, https://www.jstor.org/stable/20718325?seq=1#metadata_info_tab_contents.

18. For example, Robert Plomin, and I. J. Deary, who say that, "intelligence is one of the most heritable behavioural traits." "Genetics and Intelligence Differences: Five Special Findings," *Molecular Psychiatry* 20, no. 1 (February 2015): 98–108.

19. Sophie von Stumm and Robert Plomin, "Using DNA to Predict Intelligence," *Intelligence* 86, no. 101530 (May–June 2021); Katherine Paige Harden, *The Genetic Lottery: Why DNA Matters for Social Equality* (Princeton: Princeton University Press, 2021). For a thoughtful discussion, see Robert VerBruggen, "'The Genetic Lottery: Why DNA Matters for Social Equality'—A Review," Quillette, September 8, 2021, https://quillette.com/2021/09/08/the-genetic-lottery-why-dna-matters-for-social-equality-a-review/ .

20. Robert Plomin et al., "Nature, Nurture, and Cognitive Development from 1 to 16 Years: A Parent-Offspring Adoption Study," *Psychological Science* 8, no. 6 (November 1997): 442–47.

21. Plomin and Deary, "Genetics and Intelligence Differences."

22. Richard A.Weinberg, Sandra Scarr, and Irwin D. Waldman, "The Minnesota Transracial Adoption Study: A Follow-Up of IQ Test Performance at Adolescence," *Intelligence* 16, no. 1 (1992): 117–35.

23. From the transcript of a leaked editorial meeting at the respected monthly *The Atlantic*, May 2017. Goldberg is the editor and Coates a writer for the magazine. Ashley Feinberg, "Leak: The Atlantic Had a Meeting about Kevin Williamson. It Was a Liberal Self-Reckoning," *Huffington Post*, May 7, 2018, https://www.huffpost.com/entry/leak-the-atlantic-had-a-meeting-about-kevin-williamson-it-was-a-liberal-self-reckoning_n_5ac7a3abe4b0337ad1e7b4df.

24. Alfie Kohn, who has also criticized the idea of reward as an educational tool. Cindy Long, "Are Letter Grades Failing Our Students?," *NEA Today*, August 19, 2015, https://www.nea.org/advocating-for-change/new-from-nea/are-letter-grades-failing-our-students.

25. A surprisingly popular specialty, apparently, although Harvard is suspending their program in the wake of the pandemic. "Doctor of Educatio .

Leadership," Harvard Graduate School of Education, 2022, https://www.gse.
harvard.edu/doctorate/doctor-education-leadership.

26. Jack Schneider, "Pass-Fail Raises the Question: What's the Point of Grades?,"
New York Times, June 25, 2020, https://www.nytimes.com/2020/06/25/
opinion/coronavirus-school-grades.html. See also Philip Carl Salzman,
"Who Benefits from Cancelling Achievement Standards?," *Epoch Times*,
September 1, 2021, https://www.theepochtimes.com/who-benefits-from-
cancelling-achievement-standards_3954844.html.

27. It is hard to measure the effect on "real education" when all assessment is
abolished, of course.

28. Daniel Markovits, *The Meritocracy Trap: How America's Foundational Myth
Feeds Inequality, Dismantles the Middle Class, and Devours the Elite* (New
York: Penguin, 2019); see John Staddon, "The Meritocracy Trap—A Review,"
Quillette, October 9, 2019, https://quillette.com/2019/10/09/the-meritocracy-
trap-a-review/.

29. Markovits, *The Meritocracy Trap*, 260–61.

30. "Michael J. Sandel, Anne T. and Robert M. Bass Professor of Government,"
Harvard University, https://scholar.harvard.edu/sandel/home.

31. Michael Sandel, *The Tyranny of Merit: What's Become of the Common Good?*
(New York: Farrar, Straus and Giroux, 2020). See also Patrick Deneen's
review, which looks at Sandel from an intriguingly different angle. Patrick J.
Deneen, "A Tyranny without Tyrants?," *American Affairs* 5, no. 1 (Spring
2021): 125–39. https://americanaffairsjournal.org/2021/02/a-tyranny-without-
tyrants/.

32. This is the consensus view, but there is some dissent: Lynn O'Shaughnessy,
"The Ivy League Earnings Myth," *U.S. News and World Report*, March 1, 2011,
https://www.usnews.com/education/blogs/the-college-solution/2011/03/01/
the-ivy-league-earnings-myth.

33. Sandel, *The Tyranny of Merit*.

34. Ibid.

35. Michael J. A. Howe, "Can IQ Change?," *Psychologist* 11 (February 1998):
69–72.

36. Coaching presumably failed for the kids whose parents created the "Varsity
Blues" admissions scandal. Wikipedia, s.v. "2019 College Admissions Bribery
Scandal," https://en.wikipedia.org/wiki/2019_college_admissions_bribery_
scandal.

37. The College Board itself included. "Helping Students Practice for the SAT," College Board, 2022, https://collegereadiness.collegeboard.org/sat/k12-educators/advising-instruction/practice-resources/coaching-tools.

38. Sandel, *The Tyranny of Merit*.

39. Efforts are nevertheless being made to reconcile genetic inequality with equality of outcome. A thoughtful recent treatment is Harden's *The Genetic Lottery*.

40. Sandel, *The Tyranny of Merit*.

41. Unwillingness to believe in individual intelligence differences is becoming self-fulfilling as any kind of IQ testing falls out of favor. David Frum, "The Left's War on Gifted Kids," *The Atlantic*, July 1, 2021, https://www.theatlantic.com/politics/archive/2021/06/left-targets-testing-gifted-programs/619315/.

42. Malcolm Gladwell, *Outliers: The Story of Success* (New York: Back Bay Books, 2011).

43. Miriam A. Mosing et al., "Practice Does Not Make Perfect: No Causal Effect of Music Practice on Music Ability," *Psychological Science* 25, no. 9 (September 2014): 1795–803.

44. Michael Levin *Why Race Matters: Race Differences and What They Mean* (Westport, Connecticut: Praeger, 1997; Oakton, Virginia: New Century Foundation, 2016). The initial print run for the book was small and soon sold out. The author was savagely criticized, so the publisher apparently declined to print any more. The book was republished in 2005 by a company that has itself been attacked as racist, rendering the book unfairly tainted. Race mattered to Levin only because it had become the subject of so much legislation. The book is reviewed here: David Gordon, "The Philosopher and the IQ Debate," *Mises Review* 4, no. 4 (Winter 1998), https://mises.org/library/why-race-matters-race-differences-and-what-they-mean-michael-levin.

45. Southern Poverty Law Center, "Linda Gottfredson," https://www.splcenter.org/fighting-hate/extremist-files/individual/linda-gottfredson.

46. Linda S. Gottfredson, "Resolute Ignorance on Race and Rushton," *Personality and Individual Differences* 55, no. 3 (July 2013): 218–23.

47. Charles Murray, "An Open Letter to the Virginia Tech Community," American Enterprise Institute, March 17, 2016, https://www.aei.org/society-and-culture/an-open-letter-to-the-virginia-tech-community/.

48. Charles Lane, "The Tainted Sources of 'The Bell Curve,'" *New York Review*, December 1, 1994.

49. See, for example, the SPLC's intentionally stigmatizing "Hate Map," Southern Poverty Law Center, 2020, https://www.splcenter.org/hate-map. Their site even includes as a frequent target the heroic Ayaan Hirsi Ali.

50. Lane, "The Tainted Sources of 'The Bell Curve.'"

51. E. G. Moore, "Family Socialization and the IQ Test Performance of Traditionally and Transracially Adopted Black Children," *Developmental Psychology* 22, no. 3 (1986): 317–26. https://psycnet.apa.org/record/1986-24139-001; James R. Flynn, *Does Your Family Make You Smarter?: Nature, Nurture, and Human Autonomy* (Cambridge: Cambridge University Press, 2016).

52. "Racial Attitudes," Institute of Government and Public Affairs, University of Illinois System, https://igpa.uillinois.edu/programs/racial-attitudes; see also Ayaan Hirsi Ali, "Why Is the Southern Poverty Law Center Targeting Liberals," *New York Times*, August 24, 2017, https://www.nytimes.com/2017/08/24/opinion/southern-poverty-law-center-liberals-islam.html. For a contrary view, see Jennifer A. Richeson, "Americans Are Determined to Believe in Black Progress," *The Atlantic* (September 2020), https://www.theatlantic.com/magazine/archive/2020/09/the-mythology-of-racial-progress/614173/, which once again relies on disparities as proof of racism. On the other hand, race relations may well have deteriorated since 2011, the most recent data. John Staddon, "The Cultural Revolution: Coming to a Campus Near You," Minding the Campus, September 21, 2020, https://www.mindingthecampus.org/2020/09/21/the-cultural-revolution-coming-to-a-campus-near-you/.

53. Michael Powell, "A Tale of Two Cities," *New York Times*, May 6, 2007, https://www.nytimes.com/2007/05/06/nyregion/thecity/06hist.html.

54. Wikipedia, s.v. "Racism," last edited on June 3, 2020, http://web.archive.org/web/20200605042404/https://en.wikipedia.org/wiki/Racism.

55. Wikipedia, s.v. "Racism," last edited on February 26, 2022, https://en.wikipedia.org/wiki/Racism.

56. William MacPherson, "The Stephen Lawrence Inquiry: Report of an Inquiry by Sir William MacPherson of Cluny," February 1999, https://assets.publishing.service.gov.uk/government/uploads/system/uploads/attachment_data/file/277111/4262.pdf.

57. Ibid.

58. Ibid.

59. Zinzi D. Bailey et al., "Structural Racism and Health Inequities in the USA: Evidence and Interventions," in the series "America: Equity and Equality in

Health," *The Lancet* 389, no. 10077 (April 2017): 1453–63, https://www.
thelancet.com/journals/lancet/article/PIIS0140-6736(17)30569-X/fulltext.

60. Ibid.

61. Southern Poverty Law Center, "White Supremacists' Favorite Myths about
Black Crime Rates Take Another Hit from BJS Study," October 23, 2017,
https://www.splcenter.org/hatewatch/2017/10/23/white-supremacists-
favorite-myths-about-black-crime-rates-take-another-hit-bjs-study.

62. Ibid.

63. Rachel E. Morgan, "Race and Hispanic Origin of Victims and Offenders,
2012–15," a special report from the *Bureau of Justice Statistics*, October 2017,
https://bjs.ojp.gov/content/pub/pdf/rhovo1215.pdf.

64. Alex Berezow, "African-American Homicide Rate Nearly Quadruple the
National Average," American Council on Science and Health, August 10,
2017, https://www.acsh.org/news/2017/08/10/african-american-homicide-
rate-nearly-quadruple-national-average-11680.

65. Heather Mac Donald, "The Myth of Racial Profiling," *City Journal* (Spring
2001). See also John Staddon, "Logic of Profiling: Fairness vs. Efficiency"
(unpublished manuscript, rev. 2018), https://www.academia.edu/49043162/
The_Logic_of_Profiling_Fairness_vs_Efficiency_by.

66. Glenn Loury, "Unspeakable Truths about Racial Inequality in America,"
Quillette, February 10, 2021, https://quillette.com/2021/02/10/unspeakable-
truths-about-racial-inequality-in-america/.

67. Bailey et al., "Structural Racism and Health Inequities."

68. Kim Soffen, "How Racial Gerrymandering Deprives Black People of Political
Power," *Washington Post*, June 9, 2016, https://www.washingtonpost.com/
news/wonk/wp/2016/06/09/how-a-widespread-practice-to-politically-
empower-african-americans-might-actually-harm-them/?noredirect=on.

69. Barbara Reskin, "The Race Discrimination System," *Annual Review of
Sociology* 38 (August 2012): 18.

70. Thomas C. Schelling, "Dynamic Models of Segregation," *Journal of
Mathematical Sociology* 1, no. 2 (1971): 143–86.

71. Reskin, "The Race Discrimination System," 17.

72. Kevin Hoover, "Causality in Economics and Econometrics: An Entry for the
New Palgrave Dictionary of Economics," first draft, 2007, https://dukespace.lib.
duke.edu/dspace/bitstream/handle/10161/2046/Hoover_causality_in_
economics_and_econometrics.pdf.

73. Ben Wilterdink, "Racial Disparities and the High Cost of Low Debates,"
Quillette, May 7, 2018.

74. A recent exception is a 2021 UK report by the Commission on Race and Ethnic Disparities, which convened in response to British echoes of the 2020 BLM riots. The report concluded "that different [racial and ethnic] groups are distinguished in part by their different cultural patterns and expectations . . . [so] it is hardly shocking to suggest that some of those traditions can help individuals succeed more than others." Commission on Race and Ethnic Disparities, *Commission on Race and Ethnic Disparities: The Report* (March 2021), https://assets.publishing.service.gov.uk/government/uploads/system/uploads/attachment_data/file/974507/20210331_-_CRED_Report_-_FINAL_-_Web_Accessible.pdf. The chairman, Dr. Tony Sewell commented to the W*all Street Journal*, "In fact . . . all ethnic groups other than Caribbean blacks perform better than white British students, with the exception of Pakistanis, who are on par with whites." Tunku Varadarajan, "The Report That Shook Britain's Race Lobby," *Wall Street Journal*, April 9, 2021, https://www.wsj.com/articles/the-report-that-shook-britains-race-lobby-11617995095.

Chapter 14: Lysenko Redivivus

1. Valery N. Soyfer, "New Light on the Lysenko Era," *Nature* 339 (June 1989): 415–20.
2. For some history see David Joravsky, *The Lysenko Affair* (Chicago: University of Chicago, 1970), https://www.google.com/books/edition/The_Lysenko_Affair/nWVBgEyiiMYC?hl=en&gbpv=1.
3. See Samanth Subramanian, *A Dominant Character: The Radical Science and Restless Politics of J. B. S. Haldane* (New York: W. W. Norton, 2020).
4. See, for example, J. D. Bernal, "Stalin as Scientist," *Modern Quarterly* 8, no. 3 (Summer 1953), https://jbshaldane.org/bernal/bernal-1953-stalin-as-scientist.html.
5. Sam Kean, "The Soviet Era's Deadliest Scientist Is Regaining Popularity in Russia," *The Atlantic*, December 19, 2017, https://www.theatlantic.com/science/archive/2017/12/trofim-lysenko-soviet-union-russia/548786/.
6. Alexis de Tocqueville, *Democracy in America*, 3rd U.S. ed. , vol. 1 (New York: 1839), 259.
7. Edward Dutton in his thoughtful book *Making Sense of Race* (Georgia: Washington Summit, 2020), https://www.amazon.com/Making-Sense-Race-Edward-Dutton-ebook/dp/B08PSJ3TZ4/ref=sr_1_3?dchild=1&keywords=edward+dutton&qid=1624125218&sr=8-3, gives many examples of people who have been condemned, threatened, canceled, and even attacked, for simply

presenting data on racial differences in IQ. One may question the facts or the arguments of these people, but they should have every right to present them without fear of unpleasant consequences. It is perhaps no accident that contrarian Dutton resides in Oulu in Finland.

8. Marc Horne, "Academics Led the Campaign to Silence Genetics Professor Gregory Clark on Race," *The Times*, March 6, 2021, https://www.thetimes.co.uk/article/academics-led-the-campaign-to-silence-genetics-professor-gregory-clark-on-race-h2rrndc7w.

9. Ibid.

10. A survey is in David Reich's *Who We Are and How We Got Here: Ancient DNA and the New Science of the Human Past* (New York: Pantheon, 2018).

11. Ann Gibbons, "Busting Myths of Origin," *Science* 356, no. 6339 (May 2017): 678–81.

12. Jennifer K. Wagner et al., "Fostering Responsible Research on Ancient DNA," *American Journal of Human Genetics* 107, no. 2 (August 2020): 183–95.

13. Bruce Bourque, "The Campaign to Thwart Paleogenetic Research into North America's Indigenous Peoples," Quillette, March 29, 2021, https://quillette.com/2021/03/29/the-campaign-to-thwart-paleogenetic-research-into-north-americas-indigenous-peoples/.

14. James W. Springer and Elizabeth Weiss, "Repatriation and the Threat to Objective Knowledge," *Academic Questions* 34, no. 2 (Summer 2021), https://www.nas.org/academic-questions/34/2/repatriation-and-the-threat-to-objective-knowledge.

15. Roger Scruton, *England and the Need for Nations*, 2nd ed. (Suffolk: Saint Edmundsbury Press, 2006), 33–38, http://www.civitas.org.uk/pdf/EnglandAndTheNeedForNations.pdf.

16. Jordan G. Starck, Stacey Sinclair, and J. Nicole Shelton, "How University Diversity Rationales Inform Student Preferences and Outcomes," *Proceedings of the National Academy of Sciences* 118, no. 16 (April 2021).

17. Ibid. In two studies: "We complemented the online experiments . . . with ecologically valid samples of key university stakeholders: student caregivers (e.g., parents, grandparents, etc., $n = 255$, Study 5) and admissions officers ($n = 186$, Study 6)."

18. Ibid. White participants (details, age, etc., unspecified) were used in the first three studies; in the fourth study, participants are described as "designed to be nationally representative on the dimensions of age and gender within race."

19. "Life at Harvard: Diversity & Inclusion," Harvard College, https://college. harvard.edu/life-at-harvard/diversity-inclusion.

20. "Provost's Statement on Diversity and Inclusion," Stanford: Office of the Provost, https://provost.stanford.edu/statement-on-diversity-and-inclusion/.

21. Unfortunate terminology, implying as it does that the "educational/ instrumental" rationale for diversity is somehow "immoral."

22. Starck, Sinclair, and Shelton, "How University Diversity Rationales."

23. Acerbic satirist H. L. Mencken was wrong when he called Americans the "most timorous, sniveling, poltroonish, ignominious mob of serfs and goose-steppers ever gathered under one flag in Christendom since the end of the Middle Ages . . ." Ilana Mercer, "Rememberig H. L. Mecken: Contrarian and Gadfly," *Chronicles* (March 2020). It is not the *lumpenproletariat* who are guilty of groupthink so much as a large chunk of the intellectual classes.

24. Nicholas Wade has made a similar argument. *A Troublesome Inheritance* (New York: Penguin, 2014).

25. Philip Kitcher, "An Argument about Free Inquiry," *Nous* 31, no. 3 (September 1997): 281, referencing Stephen Jay Gould, *The Mismeasure of Man*, rev. ed. (New York: W. W. Norton, 1996). This is true enough, but ironic, given that Gould's book has been justly criticized for cherry-picking evidence and even outright falsehood: Jason E. Lewis et al., "The Mismeasure of Science: Stephen Jay Gould versus Samuel George Morton on Skulls and Bias," *PLoS Biology* 9, no. 6 (June 2011), https://journals.plos.org/plosbiology/article?id=10.1371/journal.pbio.1001071.; see also Nicholas Wade, "Scientists Measure the Accuracy of a Racism Claim," June 13, 2011, https://www.nytimes.com/2011/06/14/science/14skull.html

26. Philip Kitcher, "An Argument about Free Inquiry," 279–306.

27. Ibid.

28. See, for example, two critiques of an article by Nathan Cofnas: Rasmus Rosenberg Larson et al., "More than Provocative, Less than Scientific: A Commentary on the Editorial Decision to Publish Cofnas," *Philosophical Psychology* 33, no. 5 (August 2020): 1–6; Justin Weinberg, "Controversy at Philosophical Psychology Leads to Editor's Resignation," Daily Nous, June 24, 2020, https://dailynous.com/2020/06/24/controversy-philosophical-psychology-leads-editors-resignation/.

29. Charles Murray, *Facing Reality: Two Truths about Race in American* (New York: Encounter Books, 2021).

30. John McWhorter, "Why Charles Murray's New Book Is His Weakest . . . Despite That He Is 1) Brilliant and 2) Not a Bigot," It Bears Mentioning

(Substack), June 30, 2021, https://johnmcwhorter.substack.com/p/why-charles-murrays-new-book-is-his. Regarding Murray's information on racial IQ disparities, McWhorter notes that "It isn't that black people are on the bottom on one big test in one big study, but that a certain order of achievement manifests itself in one study after another with relentless and depressing regularity. Asians on top, then come the whites, then Latinos, and then black people."

Chapter 15: Neutral—Or Not?

1. Edward Gibbon, *History of the Decline and Fall of the Roman Empire*, rev. ed. (Cinncinati: J. A. James, 1840), 158.
2. Franklin's X-ray data (see Wikipedia, s.v. "Photo 51," https://en.wikipedia.org/wiki/Photo_51) were "stolen" by James D. Watson and formed part of the basis for his discovery, with Francis Crick, of the helical structure of DNA, all entertainingly described by Watson in his best-seller *The Double Helix: A Personal Account of the Discovery of the Structure of DNA* (London: Weidenfeld and Nicholson, 1968).
3. J. D. Bernal, *Science in History*, 3rd ed. (London: Watts & Co., 1965). See also Kenneth C. Holmes, "The Life of a Sage," *Nature* 440 (2006): 149–50.
4. Ibid., vii.
5. Ibid., 493.
6. F. A. Hayek, *The Road to Serfdom* (Chicago: University of Chicago Press, 1944), https://cdn.mises.org/Road%20to%20serfdom.pdf.
7. A process identified by Conrad Waddington in fruit flies. In a certain environment, some flies develop an abnormality. If those flies are selected and bred, eventually they will show the abnormality without the environmental challenge. "Genetic Assimilation," Fauceir Evolution, https://www.fauceir.org/wiki/page/GeneticAssimilation/.
8. J. D. Bernal, "Stalin as Scientist," *Modern Quarterly* 8, no. 3 (Summer 1953), https://jbshaldane.org/bernal/bernal-1953-stalin-as-scientist.html.
9. Bernal, *Science in History*, 469.
10. Bernal, *Science in History*, 748.
11. Bernal, *Science in History*, 749.
12. Charles Darwin to Charles Lyell, June 18, 1858, in Darwin Correspondence Project, https://www.darwinproject.ac.uk/letter/DCP-LETT-2285.xml.
13. Charles Darwin to Charles Lyell, June 25, 1858, in Darwin Correspondence Project, https://www.darwinproject.ac.uk/letter/?docId=letters/DCP-LETT-2294.xml&query=to%20publish%20a%20sketch.

14. Lytton Strachey, *Eminent Victorians* (New York: G. P. Putnam's Sons, 1918; Project Gutenberg, 2012), https://www.gutenberg.org/ebooks/2447.

15. Adrian Desmond and James Moore, *Darwin: The Life of a Tormented Evolutionist* (New York: Penguin, 1991), xx.

16. Possibly lactose intolerance. Anthony K.Campbell and Stephanie B. Matthews, "Darwin's Illness Revealed," *Postgraduate Medical Journal* 81, no. 954 (April 2005): 248–51, https://pmj.bmj.com/content/81/954/248.info.

17. Desmond and Moore, *Darwin*, xx–xxi.

18. Ibid., 552.

19. Ibid., 429. In fact, not all nineteenth-century pigeon fanciers were working class: "[The sport] was divided into two forms: short distance flying, an exclusively working-class pastime, and the more common long distance flying, a relatively expensive sport that had not insignificant numbers of middle-class followers." Martin Johnes, "Pigeon Racing and Working-Class Culture in Britain, c. 1870–1950," *Cultural and Social History* 4, no. 3 (September 2007): 361–83.

20. Charles Darwin to W. D. Fox, March 27, 1855, in Darwin Correspondence Project, https://www.darwinproject.ac.uk/letter/?docId=letters/DCP-LETT-1656.xml&query=with%20my%20outmost%20power#hit.rank1.

21. Desmond and Moore, *Darwin*, 411.

22. Ibid., 463.

23. Ibid., 413.

24. Ibid., 414.

25. Ibid., 421.

26. Ibid., 522.

27. Ibid., 463.

28. Tom Wolfe, *The Kingdom of Speech* (Boston: Little, Brown, and Company, 2016); see also John Staddon, "Open Letter to Mr. Tom Wolfe: The Kingdom of Speech: Was Charles Darwin Really an Upper-Class Bully?,"*Psychology Today*, October 11, 2016, https://www.psychologytoday.com/us/blog/adaptive-behavior/201610/open-letter-mr-tom-wolfe.

29. Desmond and Moore, *Darwin*, xix.

30. Adrian Desmond and James Moore, *Darwin's Sacred Cause: Race, Slavery, and the Quest for Human Origins* (London: Allen Lane, 2009).

31. Leon Zitzer, *A Short but Full Book on Darwin's Racism*, (Bloomington, Indiana: iUniverse, 2017).

32. A view in fact *recommended* by French "science studies" philosopher Bruno Latour. Latour received the Bernal Prize in 1992. "John Desmond Bernal

Prize," Society for Social Studies of Science, https://www.4sonline.org/what-is-4s/4s-prizes/john-desmond-bernal-prize/.

Chapter 16: Historians of Science or Political Journalists?

1. Naomi Oreskes, "The Scientific Consensus on Climate Change," *Science* 306, no. 5072 (December 2004): 1686.
2. Jill Radsken, "Defending Science in a Post-Fact Era," *Harvard Gazette*, October 22, 2019, https://news.harvard.edu/gazette/story/2019/10/in-why-trust-science-naomi-oreskes-explains-why-the-process-of-proof-is-worth-trusting/.
3. Editorial, "Science and Politics Are Inseparable," *Nature* 586 (October 2020): 169–70. For a response see Sumantra Maitra, "Who Is Responsible for the Loss of Faith in Science?," James G. Martin Center for Academic Renewal, December 11, 2020, https://www.jamesgmartin.center/2020/12/who-is-responsible-for-the-loss-of-faith-in-science/. See also Kavitha Chintam et. al., "Science Policy Can't Be Simply about Science," *Scientific American*, April 22, 2021, https://www.scientificamerican.com/article/science-policy-cant-be-simply-about-science/. Chitnam et al. argue that "the scientific community must continue working to ensure that our evolving scientific enterprise is committed to racial, economic and social equity. . . . Much of science policy remains limited to advocating for scientific research funding while neglecting broader considerations about people and communities that have been historically and repeatedly marginalized." See also Simóne D. Sun, "Stop Using Phony Science to Justify Transphobia," *Scientific American*, June 13, 2019, https://blogs.scientificamerican.com/voices/stop-using-phony-science-to-justify-transphobia/.
4. Naomi Oreskes and Erik M. Conway, *Merchants of Doubt* (London: Bloomsbury, 2010); there is also a movie *Merchants of Doubt*, directed by Robert Kenner, 2014, https://www.amazon.com/Merchants-Doubt-Patricia-Callahan/dp/B00YO2IC3W.
5. Robin McKie, "Merchants of Doubt by Naomi Oreskes and Erik M. Conway," *The Guardian* , August 7, 2010, https://www.theguardian.com/books/2010/aug/08/merchants-of-doubt-oreskes-conway.
6. Included: Frederick Seitz, Fred Singer, Richard Lindzen, and a relatively youthful Bjorn Lomborg; omitted: Will Happer, Nicola Scafetta, and Willie Soon (except as a junior author of a single cited paper).
7. Oreskes and Conway, *Merchants of Doubt*, 136.

8. Winner of the 2021 "Correlation = Causation Lazy Journalism Award" is a *Guardian* article by Damian Carrington headlined "Road [not vehicle] Pollution Affects 94% of Britain, Study Finds," *The Guardian*, March 12, 2021, https://www.theguardian.com/environment/2021/mar/12/road-pollution-affects-94-of-britain-study-finds.

9. Oreskes and Conway, *Merchants of Doubt*, 137.

10. "United States (and Tobacco-Free Kids Action Fund) v. Philip Morris, 556 F.3D 1095 (D.C. Cir. 2009)," Publice Health Law Center at Mitchell Hamline School of Law, August 8, 2014, https://www.publichealthlawcenter.org/litigation-tracker/united-states-and-tobacco-free-kids-action-fund-v-philip-morris-556-f3d-1095-dc. For a fuller account of this sorry episode in U.S. law see John Staddon, *Unlucky Strike: Private Health and the Science, Law and Politics of Smoking* (Buckingham: University of Buckingham Press, 2014).

11. Richard Doll and A. Bradford Hill, "Smoking and Carcinoma of the Lung," *British Medical Journal* 2, no. 4682 (September 1950): 739–48.

12. John Adler, "Coffin Nails: The Tobacco Controversy in the 19th Century," *Harper's Weekly*, https://tobacco.harpweek.com/hubpages/CommentaryPage.asp.

13. Bertrand Russell was born in 1872, started smoking before 1900, and was soon warned of its dangers. Bertrand Russell, "Bertrand Russell on Smoking," YouTube, https://www.youtube.com/watch?v=8ooLTiVW_lc.

14. T. Hirayama, "Non-Smoking Wives of Heavy Smokers Have a Higher Risk of Lung Cancer: A Study from Japan," *British Medical Journal (Clinical Research Edition)* 282, no. 6259 (January 1981):183–85, https://pubmed.ncbi.nlm.nih.gov/6779940/.

15. D. R. Smith and E. J. Beh, "Hirayama, Passive Smoking and Lung Cancer: 30 Years On and the Numbers Still Don't Lie," *Public Health* 125, no. 4 (April 2011):179–81, https://pubmed.ncbi.nlm.nih.gov/21661135/.

16. Hirayama, "Non-Smoking Wives," 183.

17. Oreskes and Conway, *Merchants of Doubt*, 138.

18. Hirayama, "Non-Smoking Wives," 184.

19. "Correspondence: Non-Smoking Wives of Heavy Smokers Have a Higher Risk of Lung Cancer," *British Medical Journal* 283 (October 1981): 914–15, https://www.ncbi.nlm.nih.gov/pmc/articles/PMC1507117/pdf/bmjcred00679-0044b.pdf.

20. See appendix for discussion of the replication problem created by too-generous significance levels.

21. James E. Enstrom and Geoffrey C. Kabat, "Environmental Tobacco Smoke and Tobacco Related Mortality in a Prospective Study of Californians, 1960–98," *British Medical Journal* 326, no. 7398 (May 2003):1057. https://pubmed.ncbi.nlm.nih.gov/12750205/.

22. U.S. Department of Health and Human Services, *The Health Consequences of Involuntary Exposure to Tobacco Smoke: A Report of the Surgeon General* (Atlanta, Georgia: U.S. Department of Health and Human Services, Centers for Disease Control and Prevention, Coordinating Center for Health Promotion, National Center for Chronic Disease Prevention and Health Promotion, Office on Smoking and Health, 2006), cf. 421–506, 673, https://www.ncbi.nlm.nih.gov/books/NBK44324/pdf/Bookshelf_NBK44324.pdf.

23. U.S. Department of Health and Human Services, *How Tobacco Smoke Causes Disease: The Biology and Behavioral Basis for Smoking-Attributable Disease: A Report of the Surgeon General* (Atlanta, Georgia: U.S. Department of Health and Human Services, Centers for Disease Control and Prevention, National Center for Chronic Disease Prevention and Health Promotion, Office on Smoking and Health, 2010), https://www.ncbi.nlm.nih.gov/books/NBK53017/pdf/Bookshelf_NBK53017.pdf.

24. Enstrom and Kabat, "Environment Tobacco Smoke."

25. Phillip S. Gardiner, Charles L. Gruder, and Francisco Buchting, "Rapid Response: The Case of the Footnote Wagging the Article: A Footnote with Legs," *British Medical Journal* 326 (May 2003): 1057, https://www.bmj.com/rapid-response/2011/10/29/case-footnote-wagging-article. The "footnote" refers to Enstrom's source of funding. The tobacco companies are routinely accused of funding and directing research to exonerate them from liability. In 2013 I wrote a book criticizing the legal and scientific basis for draconian anti-smoking legislation in the U.S. (published in 2014 as *Unlucky Strike*). Knowing the bad reputation of the tobacco companies, as a sort of experiment I inquired of a couple to see if they were interested in supporting the book. None were. Maybe it is a bad book? Maybe the companies thought it was a trap? Maybe they are honest?

26. Alan Rodgman, "Environmental Tobacco Smoke," *Regulatory Toxicology and Pharmacology* 16, no. 3 (December 1992): 223–44, https://www.sciencedirect.com/science/article/abs/pii/027323009290003R.

27. Kanaka Shetty et. al., "Changes in U.S. Hospitalization and Mortality Rates Following Smoking Bans" (NBER working paper no. 14790, available as SSRN), https://papers.ssrn.com/sol3/papers.cfm?abstract_id=1359506; David

Friedman, "Blowing Second-Hand Smoke," Ideas, February 6, 2013, http://daviddfriedman.blogspot.com/2013/02/blowing-second-hand-smoke.html.

28. Alan Rodgman and Thomas A. Perfetti, *The Chemical Components of Tobacco and Tobacco Smoke* (Boca Raton, Florida: CRC Press, 2009); Rodgman, "Environmental Tobacco Smoke."

29. Wikipedia, s.v. "Linear No-Threshold Model," https://en.wikipedia.org/wiki/Linear_no-threshold_model.

30. Nick Triggle, "Pub Smoking Ban: 10 Charts That Show the Impact," BBC News, July 1, 2017, https://www.bbc.com/news/health-40444460.

31. Indeed, they are explicitly ignored. Oreskes and Conway are happy to accept a *10 percent* significance level, even though current thinking about irreproducibility suggests at least the 1 percent level is necessary (if not sufficient) to ensure replicability. "Think of it this way," they write. "If you were nine-tenths sure about a crossword puzzle answer, wouldn't you write it in?" *Merchants of Doubt*, 157. There is no cost to making a mistake in a crossword puzzle. Public policy is different. Costs as well as benefits must be taken into account, but apparently not by the authors of *Merchants of Doubt*.

32. Environmental Protection Agency, "Risk Assessment for Toxic Air Pollutants: A Citizen's Guide," February 23, 2016, https://www3.epa.gov/airtoxics/3_90_024.html.

33. Oreskes and Conway, *Merchants of Doubt*, 153, 160.

34. Nassim Nicholas Taleb, *Antifragile: Things That Gain from Disorder* (New York: Random House, 2014).

35. Matt Ridley, "Dirtier Lives May Be Just the Medicine We Need," *Wall Street Journal*, September 7, 2012, https://www.wsj.com/articles/SB10000872396390443686004577633400584241864.

36. J. M. Davis and W. H. Farland, "Biological Effects of Low-Level Exposures: A Perspective from U.S. EPA Scientists," *Environmental Health Perspectives* (February 1998), https://ehp.niehs.nih.gov/doi/10.1289/ehp.98106s1379. "Mode of action" is an EPA term of art, "defined as a sequence of key events and processes, starting with interaction of an agent with a cell, proceeding through operational and anatomical changes, and resulting in cancer formation. A '*key event*' is an empirically observable precursor step that is itself a necessary element of the mode of action or is a biologically based marker for such an element. Mode of action is contrasted with '*mechanism of action*,' which implies a more detailed understanding and description of events, often at the molecular level." Risk Assessment Forum, U.S. Environment Protection Agency, *Guidelines for Carcinogen Risk Assessment*

(March 2005), chapter 1, 10, https://www.epa.gov/sites/default/files/2013-09/documents/cancer_guidelines_final_3-25-05.pdf.

37. Edward J. Calabrese, "Societal Threats from Ideologically Driven Science," *Academic Questions* 30 (Winter 2017): 405–18; https://www.nas.org/academic-questions/30/4/societal_threats_from_ideologically_driven_science; Edward J.Calabrese, "The Linear No-Threshold (LNT) Dose Response Model: A Comprehensive Assessment of Its Historical and Scientific Foundations," *Chemico-Biological Interactions* 301 (March 2019): 6–25.

38. So-called *animal models*, which seem to be valid for radiation studies but questionable for measuring the effects of many other toxins, e.g., Kahn Rhrissorrakrai et al., "Understanding the Limits of Animal Models as Predictors of Human Biology: Lessons Learned from the SBV IMPROVER Species Translation Challenge," *Bioinformatics* 31, no. 4 (February 2015): 471–83, https://academic.oup.com/bioinformatics/article/31/4/471/2748150.

39. D. Kriebel et al., "The Precautionary Principle in Environmental Science," *Environ Health Perspective* 109, no. 9 (September 2001): 871–76, https://www.ncbi.nlm.nih.gov/pmc/articles/PMC1240435/.

40. "No Safe Level of Smoking: Even Low-Intensity Smokers Are at Increased Risk of Earlier Death," National Institutes of Health, December 5, 2016, https://www.nih.gov/news-events/news-releases/no-safe-level-smoking-even-low-intensity-smokers-are-increased-risk-earlier-death.

41. EPA, "Risk Assessment for Toxic Air Pollutants."

42. Risk Assessment Forum, EPA, *Guidelines for Carcinogen Risk Assessment*.

43. Ibid. chapter 2, 14. This very odd paragraph ends by saying, "Because a latent period of up to 20 years or longer is often associated with cancer development in adults, the study should consider whether exposures occurred sufficiently long ago to produce an effect at the time the cancer is assessed. This is among the strongest criteria for an inference of causality." It is not clear whether precedence (cause comes before effect) or delay ("20 years or longer") is "among the strongest criteria."

44. Zoë Corbyn, "Naomi Oreskes: 'Discrediting Science Is a Political Strategy,'" *The Guardian*, November 3, 2019, https://www.theguardian.com/science/2019/nov/03/naomi-oreskes-interview-why-trust-science-climate-donald-trump-vaccine.

45. Nancy MacLean, *Democracy in Chains: The Deep History of the Radical Right's Stealth Plan for America* (New York: Penguin, 2017).

46. Possibly an echo of Winston Churchill's comment when he first saw his portrait, painted by Graham Sutherland and commissioned by the British House of Commons to celebrate his 80th birthday: "A remarkable example of modern art!" No mean artist himself, he hated the picture, and it was reportedly burned by his wife after his death.

47. Michael C. Munger, "On the Origins and Goals of Public Choice: Constitutional Conspiracy?" *Independent Review*, 22, no. 3 (Winter 2018): 359–82, https://www.independent.org/issues/article.asp?id=9115; Munger writes an expert 24-page critique of the science and history in the book. Veronique De Rugy, "Michael Munger: The Ultimate Nancy MaClean Slayer," *National Review*, June 30, 2017, https://www.nationalreview.com/corner/michael-munger-nancy-maclean-critique/. See also the damning retrospective by Phillip Magness, "Buchanan and the MacLean Controversy in Retrospect: 1.5 Years Later," Phillip W. Magness, http://philmagness.com/.

48. MacLean, *Democracy in Chains*.

49. Founder James Madison also saw the problem inherent in simple democracy. "A pure Democracy, by which I mean a Society consisting of a small number of citizens, who assemble and administer the Government in person, can admit of no cure for the mischiefs of faction." *The Federalist Papers*, no. 10. In other words, a majority faction should not be omnipotent.

50. Howard Zinn, *A People's History of the United States* (New York: HarperCollins, 2017). See Mary Grabar's *Debunking Howard Zinn: Exposing the Fake History That Turned a Generation against America* (Washington, D.C.: Regnery History, 2019).

51. Nikole Hannah-Jones, *The 1619 Project, New York Times Magazine*, August 14, 2019; Editor in Chief, "We Respond to the Historians Who Critiqued the 1619 Project: Five Historians Wrote to Us with Their Reservations, Our Editor in Chiefs Replies," *New York Times*, January 19, 2021, https://www.nytimes.com/2019/12/20/magazine/we-respond-to-the-historians-who-critiqued-the-1619-project.html.

52. James Buchanan and Gordon Tullock, "The Conceptual Framework," in the landmark work *The Calculus of Consent: Logical Foundations of Constitutional Democracy* (Indianapolis, Indiana: Liberty Fund, 1999).

53. Eminent postmodern philosopher in Duke's Literature program; also 1997 winner of the late Dennis Dutton's famed International Bad Writing Contest, "The Bad Writing Contest," Press Releases, 1996–1998, http://www.denisdutton.com/bad_writing.htm.

54. Munger, "On the Origins and Goals."

55. MacLean, *Democracy in Chains*. For more on the Kochs' noninfluence on Buchanan, in addition to the Munger review, see Brian Doherty, "What Nancy MacLean Gets Wrong About James Buchanan," *Reason*, July 20, 2017.

56. James M. Buchanan Jr., "Prize Lecture," December 8, 1986, https://www.nobelprize.org/prizes/economic-sciences/1986/buchanan/lecture/.

57. Tullock even extended his work to behavioral ecology, viz the wonderfully titled "The Coal Tit as a Careful Shopper," *American Naturalist* 105, no. 941 (January–February 1971): 77–80.

58. Buchanan and Tullock, *The Calculus of Consent*, 6.

59. Sam Tanenhaus, "The Architect of the Radical Right: How the Nobel Prize–Winning Economist James M. Buchanan Shaped Today's Antigovernment Politics," *The Atlantic* (July/August 2017).

60. Oreskes and MacLean are not alone. Rebecca Lemov, another Harvard historian, has followed the same path in her account of *behaviorism*: see review by John E. Staddon, "The Behaviorist Plot," *Academic Questions* 34, no. 2 (Summer 2021), https://www.nas.org/academic-questions/34/2/the-behaviorist-plot.

Epilogue

1. Jeffery K. Taubenberger and David M. Morens, "1918 Influenza: The Mother of All Pandemics," *Emerging Infectious Diseases* 12, no. 1 (January 2006), https://www.ncbi.nlm.nih.gov/pmc/articles/PMC3291398/.

2. Polly J. Price, "How a Fragmented Country Fights a Pandemic," *The Atlantic*, March 19, 2020, https://www.theatlantic.com/ideas/archive/2020/03/how-fragmented-country-fights-pandemic/608284/.

3. David Adam, "Special Report: The Simulations Driving the World's Response to COVID-19," *Nature*, April 3, 2020, https://www.nature.com/articles/d41586-020-01003-6.

4. Peter St. Onge and Gaël Campan, "The Flawed COVID-19 Model That Locked Down Canada," Montreal Economic Institute, June 4, 2020, https://www.iedm.org/the-flawed-covid-19-model-that-locked-down-canada/.

5. Andrew Scheuber and Sabine L. van Elsland, "BMJ Study Confirms Imperial COVID-19 Projections," Imperial College London, October 8, 2020, https://www.imperial.ac.uk/news/206213/bmj-study-confirms-imperial-covid-19-projections/.

6. As of March 2021. Centers for Disease Control and Prevention, "COVID-19 Forecasts: Deaths," March 3 2021, https://web.archive.org/

web/20210308221324/https://www.cdc.gov/coronavirus/2019-ncov/covid-data/forecasting-us.html.

7. Ibid.

8. Adam, "Special Report."

9. For a partial treatment, see John Staddon, *Scientific Method: How Science Works, Fails to Work or Pretends to Work*, (London: Taylor and Francis, 2017), chapters 5–7; and John Staddon, *The Malign Hand of the Markets: The Insidious Forces on Wall Street That Are Destroying Financial Markets—And What We Can Do about It* (New York: McGraw Hill, 2012).

10. *Best of Enemies*, directed by Morgan Neville and Robert Gordon, featuring William F. Buckley and Gore Vidal, 2015, https://www.amazon.com/Best-Enemies-William-F-Buckley/dp/B01771PI4M.

11. Simona Varrella, "Homosexuality in the United States—Statistics & Facts," Statista, May 31, 2021, https://www.statista.com/topics/1249/homosexuality/. There appears to be a good biological reason for the male-female statistical discrepancy. Evolution favors variability in the male more than variability in the female. It pays males to take a chance; females, not so much. Wikipedia, s.v. "Variability Hypothesis," https://en.wikipedia.org/wiki/Variability_hypothesis.

12. Scott Jaschik, "What Larry Summers Said," Inside Higher Education, February 18, 2005, https://www.insidehighered.com/news/2005/02/18/what-larry-summers-said. It is sad to relate the reaction of one listener to Summers's speech. Distinguished biologist Nancy Hopkins, then at MIT, seemed to confirm common stereotypes—women more emotional and less super-smart than men, etc.—when she said later, "If I hadn't left the room, I would've either blacked out or thrown up" (when Summers made the [valid] point about male vs. female variability). "Soundbites," *NewScientist*, January 19, 2005.

13. John Staddon, "Values, Even Secular Ones, Depend on Faith: A Reply to Jerry Coyne," Quillette, April 28, 2019, https://quillette.com/2019/04/28/values-even-secular-ones-depend-on-faith-a-reply-to-jerry-coyne/.

14. Bret Stephens, "Read the Column the *New York Times* Didn't Want You to See," New York Post, February 11, 2021, https://nypost.com/2021/02/11/read-the-column-the-new-york-times-didnt-want-you-read/.

Appendix: *The Replication Crisis and its Offspring*

1. Although the use of NHST statistics in psychophysical research has increased since the 1960, with no discernible benefit: Scott Parker, "A Note on the

Growth of the Use of Statistical Tests in *Perception and Psychophysics*," *Bulletin of the Psychonomic Society* 28, no. 6 (December 1990): 565–66.

2. *Situation* means stimuli present—key color, Skinner box, etc.—and associated reward schedule. Recovery of a learned performance when the stimulus situation recurs is termed *stimulus control*. For details see (for example) John Staddon, *Adaptive Behavior and Learning*, 2nd ed. (Cambridge: Cambridge University Press, 2016).

3. Daryl J. Bem, "Feeling the Future: Experimental Evidence for Anomalous Retroactive Influences on Cognition and Affect," *Journal of Personality and Social Psychology* 100, no. 3 (March 2011): 407–25. Bem's ambitious experiment (he appeared, appropriately enough, on the Colbert Report, "The Colbert Report—Daryl Bem," March 29, 2019, https://econ.video/2019/03/29/the-colbert-report-daryl-bem/) has been admirably dispatched by Stuart Ritchie in his book *Science Fictions: How Fraud, Bias, Negligence, and Hype Undermine the Search for Truth* (New York: Metropolitan Books, 2020).

4. Benjamin U. Phillips et al., "Optimisation of Cognitive Performance in Rodent Operant (Touchscreen) Testing: Evaluation and Effects of Reinforcer Strength," *Learning & Behavior* 45, no. 3 (September 2017): 252–62. The rewards were strawberry milkshake and something called super saccharine, plus a couple of others.

5. John P. A. Ioannidis, "Why Most Published Research Findings Are False," *PLoS Medicine* 2, no. 8 (August 2005), https://www.ncbi.nlm.nih.gov/pmc/articles/PMC1182327/; see also "Reproducibility and Replicability in Science," National Academies of Sciences, https://www8.nationalacademies.org/pa/projectview.aspx?key=49906.

6. See, for example, J. Cohen, "A Power Primer," *Psychological Bulletin*, 112, no. 1 (July 1992): 155–59; Robert Coe, "It's the Effect Size, Stupid: What Effect Size Is and Why It Is Important," (conference paper, Annual Conference of British Educational Research, Exeter, England, September 2002), https://www.semanticscholar.org/paper/It-%27-s-the-Effect-Size-%2C-Stupid-What-effect-size-is-Coe/c5ac87df5d6e0e6b6de2f745284835c2a368b0f7.

7. Christie Aschwanden and Ritchie King, "Science Isn't Broken: It's Just a Hell of a Lot Harder than We Give It Credit for," FiveThirtyEight, August 19, 2014, https://fivethirtyeight.com/features/science-isnt-broken/#part1.

8. Wikipedia, s.v. "Data Dredging," https://en.wikipedia.org/wiki/Data_dredging.

9. Chris Chambers, *The Seven Deadly Sins of Psychology: A Manifesto for Reforming the Culture of Scientific Practice* (Princeton: Princeton University

Press, 2017); David Randall and Christopher Welser, "The Irreproducibility of Modern Science: Causes, Consequences, and the Road to Reform," National Association of Scholars, 2018, https://www.nas.org/reports/the-irreproducibility-crisis-of-modern-science.

10. For example, the Association for Psychological Science's "Preregistration of Research Plans," https://www.psychologicalscience.org/publications/psychological_science/preregistration.

11. For a review, see Kevin Hoover, "Causality in Economics and Econometrics: An Entry for the *New Palgrave Dictionary of Economics*," first draft, 2007, https://dukespace.lib.duke.edu/dspace/bitstream/handle/10161/2046/Hoover_causality_in_economics_and_econometrics.pdf.

12. Journalists often use the term *linked*, which seems to imply causation without actually saying so.

13. Ronald A. Fisher, "Cigarettes, Cancer, and Statistics," *Centennial Review of Arts & Science* 2 (1958): 151–61.

14. Daniel J. Benjamin, "Redefine Statistical Significance," *PsyArXiv*, July 22, 2017; last edited February 3, 2020, https://psyarxiv.com/mky9j/, described in Kelly Servick, "It Will Be Much Harder to Call New Findings 'Significant' If This Team Gets Its Way," *Science*, July 25, 2017, https://www.sciencemag.org/news/2017/07/it-will-be-much-harder-call-new-findings-significant-if-team-gets-its-way.

15. There are many excellent online accounts of p-value (alpha) criteria and the two types of error in NHST statistics, e.g., Saul McLeod, "What Are Type I and Type II Errors," Simply Psychology, July 4, 2019, https://www.simplypsychology.org/type_I_and_type_II_errors.html.

16. But not always: Brian Owens, "Replication Failures in Psychology Not Due to Differences in Study Populations," *Nature*, November 19, 2018, https://www.nature.com/articles/d41586-018-07474-y.

17. Daniel Lakens et al., "Justify Your Alpha," *Nature Human Behavior* 2 (February 2018): 168–71, https://www.nature.com/articles/s41562-018-0311-x; Daniel Lakens et al., "Justify Your Alpha," *PsyArXiv*, September 18, 2017, https://psyarxiv.com/9s3y6/.

18. For reviews, see Monya Baker, "1,500 Scientists Lift the Lid on Reproducibility," *Nature* 533 (May 2016); 452–54; "Reproducibility and Replicability in Science," National Academies of Sciences; and John Staddon, *Scientific Method: How Science Works, Fails to Work or Pretends to Work* (London: Taylor and Francis, 2017).

19. R. A. Fisher, *Statistical Methods for Research Workers*, 5th ed., vol. 5 of *Biological Monographs and Manuals*, eds. F. A. E. Crew and D. Ward Cutler (Edinburgh: Oliver and Boyd, 1934), http://www.haghish.com/resources/materials/Statistical_Methods_for_Research_Workers.pdf.

20. False positives, just like scientific frauds, can have very damaging effects: Kai Kupferschmidt, "Researcher at the Center of an Epic Fraud Remains an Enigma to Those Who Exposed Him," *Science*, August 17, 2018, https://www.science.org/content/article/researcher-center-epic-fraud-remains-enigma-those-who-exposed-him; Theodore Dalrymple, *False Positive: A Year of Error, Omission and Political Correctness in the New England Journal of Medicine* (New York: Encounter Books, 2019); Ritchie describes many other cases in *Science Fictions*. See also Andrew Gelman, "A Scandal in Tedhemia: Noted Study in Psychology First Fails to Replicate (But Is Still Promoted by NPR), Then Crumbles with Striking Evidence of Data Fraud," Statistical Modeling, Causal Inference, and Social Science, August 19, 2021, https://statmodeling.stat.columbia.edu/2021/08/19/a-scandal-in-tedhemia-noted-study-in-psychology-first-fails-to-replicate-but-is-still-promoted-by-npr-then-crumbles-with-striking-evidence-of-data-fraud/.

21. S. Stanley Young, Warren Kindzierski, and David Randall, *Shifting Sands: Unsound Science and Unsafe Regulation* (New York: National Association of Scholars, 2021), https://www.nas.org/reports/shifting-sands-report-i/full-report#_ftnref76; Jesse Singal, "Positive Psychology Goes to War: How the Army Adopted an Untested, Evidence-Free Approach to Fighting PTSD," *Chronicle of Higher Education*, June 7, 2021, https://www.chronicle.com/article/positive-psychology-goes-to-war.

22. For more on this topic, see John Staddon, *The New Behaviorism: Foundations of Behavioral Science*, 3rd ed. (Philadelphia: Psychology Press, 2021).

23. Claude Bernard, *An Introduction to the Study of Experimental Medicine [Introduction à l'étude de la médecine expérimentale]* (1865) Kindle, 1255–56. Translation of "on ne devra jamais admettre des exceptions ni des contradictions réelles, ce qui serait antiscientifique," and many similar sentences in *Experimental Medicine*.

24. If a survey is repeated, the results from different samples can be compared to get a *margin of error* for the overall average. But usually, the size of the single sample is used as an adequate measure of margin of error.

25. *Expected gain* = p (outcome) x amount, in this case 0.2 x 4000 = 800.

26. Daniel Kahneman and Amos Tversky, "Prospect Theory: An Analysis of Decision under Risk," *Econometrica* 47, no. 2 (March 1979): 263–91.

27. Wikipedia, s.v. "List of Cognitive Biases," https://en.wikipedia.org/wiki/List_of_cognitive_biases.

28. Kahneman and Tversky, "Prospect Theory," figure 3. This is a teleological, not causal, account with all the limitations I described in *The New Behaviorism*, chapter 8. Kahneman and Tversky published a more complex, but still not causal, version of prospect theory in 1992. Amos Tversky and Daniel Kahneman, "Advances in Prospect Theory: Cumulative Representation of Uncertainty," *Journal of Risk and Uncertainty* 5, no. 4 (1992): 297–322.

29. See Jason Collins, "Gigerenzer versus Kahneman and Tversky: The 1996 Face-Off," Jason Collins Blog, April 1, 2019, https://www.jasoncollins.blog/gigerenzer-versus-kahneman-and-tversky-the-1996-face-off/, for a discussion of the debate between Kahneman and Tversky and their major critic, Gerd Gigerenzer.

30. See Wikipedia, s.v. "Allais Paradox," https://en.wikipedia.org/wiki/Allais_paradox.

Index

and ideology, xiii–xiv, 42, 47, 118, 174–82

and morality, 7–13

nature of, xi–xii, xv, 1, 21, 123–25, 167, 207

and peer review, 57–70

politicization of, xii–xiv, 24, 43, 46, 73, 95, 115, 134–35, 161, 177, 182–83, 186, 192, 204, 206–7

as a profession, xiii, 37–47, 58–59

publication of, 49–55, 57–70

whether a religion, 1–6

social (*see* social science)

Scruton, Roger, 134, 163

secondhand smoke, 184–86, 188

secular humanism, 1–6, 207, 227n17

segregation, racial, 127, 151, 199

Skinner, B. F., 3, 15–16, 20–21, 231n9

Smith, Adam, 198, 204

smoking, 17–18, 21, 23, 184–94, 215

social facts, 112, 117, 121

social justice, xiii, 44, 47, 164–65, 183

social pressure, 160, 167

social science, xii–xiv, 40, 42, 45, 54, 73, 99, 101–3, 107–15, 118, 121, 123–24, 127, 131, 133–35, 152, 156, 161, 174, 176, 198, 216

sociology, xii, 107, 110–12, 115, 117–135, 137, 156–57

species, 28, 94, 180, 205–6

constancy of, 27

survival of, 19–21

evolution of, 16, 25, 27, 31–32

Stalin, Josef, 159–60, 176

statistical significance, 212, 217

STEM, 38, 43, 45, 102–3, 165, 233n5, 236n25

structural racism. *See* systemic racism

systemic racism, xiv, 125–26, 130, 137–58, 170

systems theory, 155–58

T

temperature, 80–83, 85–96, 242n21

tenure, 37–38, 68

Tocqueville, Alexis de, 159–60, 166

trans-science, xii–xiv, 73, 99–100, 141, 185,

Tversky, Amos, 219–22

U

uniformitarianism, 29

V

variables

dependent, 91, 212–13

independent, 212–13

voter suppression, 155